Springer INdAM Series

Volume 46

This series will publish textbooks, multi-authors books, thesis and monographs in English language resulting from workshops, conferences, courses, schools, seminars, doctoral thesis, and research activities carried out at INDAM - Istituto Nazionale di Alta Matematica, http://www.altamatematica.it/en. The books in the series will discuss recent results and analyze new trends in mathematics and its applications.

THE SERIES IS INDEXED IN SCOPUS

More information about this series at http://www.springer.com/series/10283

Vincenzo Vespri • Ugo Gianazza •
Dario Daniele Monticelli • Fabio Punzo •
Daniele Andreucci
Editors

Harnack Inequalities and Nonlinear Operators

Proceedings of the INdAM conference to celebrate the 70th birthday of Emmanuele DiBenedetto

Springer

Editors
Vincenzo Vespri
Dipartimento di Matematica e Informatica
Università di Firenze
Firenze, Italy

Ugo Gianazza
Dipartimento di Matematica
Università di Pavia
Pavia, Italy

Dario Daniele Monticelli
Dipartimento di Matematica
Politecnico di Milano
Milano, Italy

Fabio Punzo
Dipartimento di Matematica
Politecnico di Milano
Milano, Italy

Daniele Andreucci
Dipartimento di Scienze di Base e Applicate
per l'Ingegneria
Università di Roma "La Sapienza"
Rome, Italy

ISSN 2281-518X ISSN 2281-5198 (electronic)
Springer INdAM Series
ISBN 978-3-030-73777-1 ISBN 978-3-030-73778-8 (eBook)
https://doi.org/10.1007/978-3-030-73778-8

This Springer imprint is published by the registered company Springer Nature Switzerland AG.
The registered company address is: Gewerbestrasse 11, 6330 Cham, Switzerland

Emmanuele DiBenedetto
April 2nd 1947, Lentini (Italy) – May 11th 2021, Nashville (TN-USA)

Contents

What I Learnt from Emmanuele DiBenedetto

Vincenzo Vespri

Abstract Emmanuele was neither my Master thesis supervisor, nor my PhD advisor, but, nevertheless, it was a turning point in my life, both scientifically and personally. In this note I will describe how I decided to work with him on the regularity estimates for weak solutions to degenerate/singular parabolic equations an d on Harnack estimates More precisely, here I will give an outline of the proof of the regularity estimates and Harnack's estimates only for weak solutions to degenerate p-Laplacean equations (i.e. for $p > 2$). For the sake of simplicity, I will prove these results only for the prototype parabolic equations but the techniques exposed here work also in a very general setting.

Keywords Degenerate and parabolic equations · Regularity and Harnack estimates · Di Benedetto

1 Introduction

Emmanuele was neither my Master thesis supervisor, nor my PhD advisor, but, nevertheless, it was a turning point in my life, both scientifically and personally. The first time I heard of him, was when I was a ricercatore (the corresponding position of Assistant Professor in US) at the Second University of Rome in Tor Vergata. I was quite young and I was trying to define my own line of scientific research. In the meanwhile, Emmanuele, a recognized Math superstar was coming from the US to teach in my same university: the occasion was as propitious as possible. Then things did not turn out that way: his arrival in Tor Vergata was slowed down by academic fights (in Italy we use the expression baronial fights) and, in the meantime, I was hired as Associate Professor at the University of Milan. Nevertheless, the decision to be his student had been made and, after few months, I ended up being a visiting

V. Vespri (✉)
Dipartimento di Matematica ed Informatica, Università di Firenze, Florence, Italy
e-mail: vincenzo.vespri@unifi.it

V. Vespri et al. (eds.), *Harnack Inequalities and Nonlinear Operators*,
Springer INdAM Series 46, https://doi.org/10.1007/978-3-030-73778-8_1

Professor for a year at Northwestern University in Evanston (where at that time Emmanuele had the main academic position). And now, it is 30 years that I have been working on his research arguments.

What can I say about Emmanuele? Certainly, he is an excellent mathematician, but also a very tough guy, both with others, but above all with himself. His not easy academic life made him accustomed to fight and survive in hostile environments. At work, he is very professional. He is open (as a Sicilian can be) only when is not working, i.e. when he is dining, jogging, taking a sauna, etc. Therefore, these are the best moments to discuss Mathematics, and not only that.

On these occasions he is used to dispense thoughts, reflecting his ideas about life. Here, I collect only a few.

- In academy, one can advance in his career, either by licking asses or by working hard. But if one works hard, then one not only preserves his dignity, but has the great advantage to be free, because nobody can ask you for favors to be returned.
- Do not put trust in the others. One can be generous for oneself, but one must not count on the gratitude of those who were helped. In the real life, one can only ask for what one can taken by brute force.
- A mathematician is remembered thanks to his theorems, not for his university political activity, which is only vanitas vanitatum. For this reason, one must only devote himself to significant problems. Simple problems (he calls them using the French word *alimentaire* to joke with the French word *elementaire*) must be left to young people who have to cut their teeth.
- Mathematics is frustrating. It is like a soccer game. Only a few goals are scored in one game. There are (too) many moments when you cannot score. To overcome these moments, it is better to do Mathematics with friends. Mathematics (together with Philosophy) is a convivial science. To be discussed at the table. But it must be clear that Mathematics requires a lot of efforts, a lot of calculations to do, when one is sitting at the desk. The deepest Theorems are based not only on good ideas but also on "sweaty papers".

Also, as an Italian coming from Sicily, who emigrated to the US (too) many years ago, the relationship with his homeland is not easy, based on a hate-love sentiment. It is evident that he feels very much in debt with Italy: for the Math education he received, for the Italian culture he learned when was a young student in Catania and in Florence. Under this point of view, one can understand his decision to return to teach in Italy for a decade, and why many of his "disciples" are Italian. But, at the same time, he is very critical with his own country: he clearly sees all the flaws that prevent Italy from being a normal country (using his words).

Fig. 1 Emmanuele at a conference

He is a wonderful lecturer and teacher. When he teaches, he is very charismatic and very tough. I met some of his undergraduate students some years after their graduation. Even if they passed his exam with low marks, all of them feel almost a sentiment of veneration for his way to teach.

The main contributions made by Emmanuele are the ones concerning the regularity of solutions to the p-Laplacean equations (for his results, Neil Trudinger named Emmanuele the p-Laplacean man) and the Harnack inequalities for degenerate parabolic quasilinear equations. For this last achievements, Juan Luis Vazquez named him and his students (Ugo and me) as the Harnack's brothers.

In this note, I will give an outline of the proof of the regularity estimates and Harnack's estimates only for weak solutions to degenerate p-Laplacean equations (i.e. for $p > 2$). For a more complete exposition of these results, I refer the reader to the monographs [10] and [16]. More precisely, first, I will present a modified proof of the regularity of the solution to linear parabolic equations. Then, starting from this proof, I will consider degenerate p-Laplacean equations and then I will prove the Hölder regularity and the Harnack estimates of the solutions. For the sake of simplicity, I will prove these results for the prototype parabolic equations but the techniques exposed here work also in a very general setting (Fig. 1).

2 Linear Equations: Preliminary Results

Let Ω be an open bounded set in $\mathbb{R}^N$, let $T > 0$ and define $\Omega_T \equiv \Omega \times (0, T)$. Consider a local weak solution of the heat equation

$$u_t = \Delta u. \tag{2.1}$$

in Ω_T, that is $u \in L^2_{loc}(0, T); W^{1,2}_{loc}(\Omega)) \cap C_{loc}((0, T); L^2_{loc}(\Omega))$ such that for any function ϕ with compact support in Ω_T and regular enough, we have that for any $0 < s < t < T$

$$\int_s^t \int_\Omega (u\phi_t + \sum_{i=1}^N D_i u D_i \phi) dx dt = \int_\Omega u(x, t)\phi(x, t) dx - \int_\Omega u(x, s)\phi(x, s) dx. \tag{2.2}$$

Here we give a proof that any bounded solution is Hölder continuous. The result is classic and goes back to the classical Schauder estimates (see [34, 35]). See also the monograph by Gilbarg and Trudinger [22]. Here we will give a proof that works also for bounded weak solutions to equations with only measurable coefficients, i.e.

$$u_t = \sum_{i=1}^N D_i(a_{i,j} D_j u)$$

with $a_{i,j}$ locally bounded, measurable and elliptic, i.e. there is a $\nu > 0$ such that for any $\xi \in \mathbb{R}^N$ we have $\sum_{i=1}^N a_{i,j}\xi_i\xi_j \geq \nu\|\xi\|$. Such a result was proved by Nash [32] directly for solution to parabolic equations and by De Giorgi [5] for solution to the corresponding elliptic equations. The technique introduced by De Giorgi was extended by Ladyzhenskaja, Solonnikov and Uralceva to the parabolic setting in [26]). Here we follow the De Giorgi approach, because it turned out to be very flexible and suitable to be extended to the nonlinear case. We recall also that using an interation technique introduced by De Giorgi and Moser (see [5, 29], see also [26]), it is possible to prove that any weak solution is bounded, but, here, for the sake of simplicity, we will not prove such an important result.

Let us start recalling that the De Giorgi approach to prove the continuity of solutions at a point consists, roughly speaking, in constructing a sequence of nested and shrinking cylinders with vertex at that point, in a way that the essential oscillation of the function in those cylinders converges to zero when the cylinders shrink to zero. The proof is based on energy (also known as Caccioppoli estimates) and logarithmic estimates for the solutions. In other words, quite surprisingly, the De Giorgi approach says that any bounded function $u \in L^2_{loc}((0, T); W^{1,2}_{loc}(\Omega)) \cap C_{loc}((0, T); L^2_{loc}(\Omega))$ satisfying the energy and logarithmic estimates is also α-Hölder continuous for some $\alpha > 0$. This means that the Hölder continuity is a property of a class of functions (the so called De Giorgi classes) more than a property of being a solution to a partial differential equation. The only thing to do is proving that a local solution to (2.1) satisfies also the energy and logarithmic estimates.

2.1 Local Energy and Logarithmic Estimates

Given a point $x_0 \in \mathbb{R}^N$, let us denote by $K_\rho(x_0)$ the N-dimensional cube with centre at x_0 and wedge 2ρ:

$$K_\rho(x_0) := \left\{ x \in \mathbb{R}^N : \max_{1 \le i \le N} |x_i - x_{0i}| < \rho \right\} ;$$

given a point $(x_0, t_0) \in \mathbb{R}^{N+1}$, the cylinder of radius ρ and height $\tau > 0$ is

$$(x_0, t_0) + Q(\tau, \rho) := K_\rho(x_0) \times (t_0 - \tau, t_0) .$$

Consider a cylinder $(x_0, t_0) + Q(\tau, \rho) \subset \Omega_T$ and let $0 \le \zeta \le 1$ be a piecewise smooth cutoff function in $(x_0, t_0) + Q(\tau, \rho)$ such that

$$|\nabla \zeta| < \infty \quad \text{and} \quad \zeta(x, t) = 0 , \quad x \notin K_\rho(x_0) . \tag{2.3}$$

Let us start with the energy estimates. Without loss of generality, we will state them for cylinders with "vertex" at the origin $(0, 0)$.

Lemma 2.1 (Energy Estimates) *Let u be a local weak solution of (2.1). There exists a constant $C > 0$ such that for every cylinder $Q(\tau, \rho) \subset \Omega_T$,*

$$\sup_{-\tau < t < 0} \int_{K_\rho \times \{t\}} (u - k)_\pm^2 \, \zeta^2 \, dx + \int_{-\tau}^{0} \int_{K_\rho} |\nabla (u - k)_\pm \zeta|^2 \, dx \, dt$$

$$\le \int_{K_\rho \times \{-\tau\}} (u - k)_\pm^2 \, \zeta^2 \, dx + C \int_{-\tau}^{0} \int_{K_\rho} (u - k)_\pm^2 \, (|\nabla \zeta|^2 + 2\zeta |\partial_t \zeta|) \, dx \, dt. \tag{2.4}$$

Proof Use $\varphi = \pm(u_h - k)_\pm \zeta^2$ as a testing function in (2.2) and perform standard calculations (see [10, pages 24–27]).

Let us state the logarithmic estimates. Given constants a, b, c, with $0 < c < a$, define the nonnegative function

$$\psi^{\pm}_{\{a,b,c\}}(s) \equiv \left(\ln \left\{ \frac{a}{(a + c) - (s - b)_\pm} \right\} \right)_+$$

$$= \begin{cases} \ln \left\{ \frac{a}{(a+c) \pm (b - s)} \right\} & \text{if } b \pm c \;{}_{>}^{<}\; s \;{}_{>}^{<}\; b \pm (a + c) \\ 0 & \text{if } s \;{}_{\ge}^{\le}\; b \pm c \end{cases}$$

whose first derivative is

$$\left(\psi^{\pm}_{\{a,b,c\}}\right)'(s) = \frac{a}{(a+c) \pm (b-s)} \quad \text{if} \quad b \pm c \,{}_{>}^{<}\, s \,{}_{>}^{<}\, b \pm (a+c) \,{}_{\leq}^{\geq}\, 0\,,$$

and

$$\left(\psi^{\pm}_{\{a,b,c\}}\right)'(s) = 0 \quad \text{if} \quad s \,{}_{>}^{<}\, b \pm c.$$

The second derivative, with $s = b \pm c$, is

$$\left(\psi^{\pm}_{\{a,b,c\}}\right)'' = \left\{\left(\psi^{\pm}_{\{a,b,c\}}\right)'\right\}^2 \geq 0.$$

Let u be a bounded function defined in a cylinder $(x_0, t_0) + Q(\tau, \rho)$ and let k be a number. Define the constant

$$H^{\pm}_{u,k} \equiv \operatorname*{ess\,sup}_{(x_0,t_0)+Q(\tau,\rho)} |(u-k)_{\pm}|\,.$$

Let

$$\Psi^{\pm}\left(H^{\pm}_{u,k}, (u-k)_{\pm}, c\right) \equiv \psi^{\pm}_{\left\{H^{\pm}_{u,k},k,c\right\}}(u)\,, \quad 0 < c < H^{\pm}_{u,k}\,.$$

This function was introduced in [26] and extended to the p-Laplacean case in [6] and it is a fundamental tool for proving regularity results for degenerate PDE's. Let $x \mapsto \zeta(x)$ be a time-independent cutoff function in $K_\rho(x_0)$ satisfying (2.3). Now, we are ready to state the logarithmic estimates. □

Lemma 2.2 (Logarithmic Estimates) *Let u be a local weak solution of (2.1), $k \in \mathbb{R}$ and $0 < c < H^{\pm}_{u,k}$. There exists a constant $C > 0$ such that for every cylinder $Q(\tau,\rho) \subset \Omega_T$,*

$$\sup_{-\tau<t<0} \int_{K_\rho \times \{t\}} \left[\psi^{\pm}(u)\right]^2 \zeta^2 \,\mathrm{d}x \leq \int_{K_\rho \times \{-\tau\}} \left[\psi^{\pm}(u)\right]^2 \zeta^2 \,\mathrm{d}x$$
$$+ C \int_{-\tau}^{0} \int_{K_\rho} \psi^{\pm}(u)\, |\nabla \zeta|^2 \,\mathrm{d}x\,\mathrm{d}t\,. \tag{2.5}$$

Proof Take $\varphi = 2\,\psi^{\pm}(u_h)\left[(\psi^{\pm})'(u_h)\right]\zeta^2$ as a testing function in (2.2) and integrate in time over $(-\tau, t)$ for $t \in (-\tau, 0)$. The result comes from nowadays standard calculations (see [13, pages 181–183]). □

2.2 *Preliminary Technical Results*

Here we collect some technical facts that, although marginal to the theory, are essential in this approach to the regularity theory.

Given a continuous function $v : \Omega \to \mathbf{R}$ and two real numbers $k < l$, we define

$$\begin{aligned} [v > l] &\equiv \{x \in \Omega \ : \ v(x) > l\} \ , \\ [v < k] &\equiv \{x \in \Omega \ : \ v(x) < k\} \ , \\ [k < v < l] &\equiv \{x \in \Omega \ : \ k < v(x) < l\} \ . \end{aligned} \tag{2.6}$$

Lemma 2.3 (De Giorgi, [5]) *Let $v \in W^{1,1}\left(B_\rho(x_0)\right)$, with $\rho > 0$ and $x_0 \in \mathbf{R}^N$ and $k < l \in \mathbf{R}$. There exists a constant C, depending only on N and p (so independent of ρ, x_0, v, k and l), such that*

$$(l-k)\ |[v > l]| \leq C\,\frac{\rho^{N+1}}{|[v<k]|}\int_{[k<v<l]} |\nabla v|\, \mathrm{d}x\ .$$

The following lemma concerns the geometric convergence of sequences.

Lemma 2.4 (Geometric Convergence, [5]) *Let $\{X_n\}$, $n = 0, 1, 2, \ldots$, be a sequence of positive real numbers satisfying the recurrence relation*

$$X_{n+1} \leq C\, b^n\, X_n^{1+\alpha}$$

where $C, b > 1$ and $\alpha > 0$ are given. If

$$X_0 \leq C^{-1/\alpha}\, b^{-1/\alpha^2}$$

then $X_n \to 0$ as $n \to \infty$.

Let $V_0^2(\Omega_T)$ denote the space

$$V_0^2(\Omega_T) = L^\infty\left(0, T; L^2(\Omega)\right) \cap L^2\left(0, T; W_0^{1,2}(\Omega)\right)$$

endowed with the norm

$$\|u\|^2_{V^2(\Omega_T)} = \ \text{ess}\sup_{0\leq t\leq T} \|u(\cdot, t)\|^2_{2,\Omega} + \|\nabla u\|^2_{2,\Omega_T}\ ,$$

for which the following embedding theorem holds (see [10, page 9]):

Theorem 2.5 (Sobolev Embeddings) *There exists a constant γ, depending only upon N, such that for every $v \in V_0^2(\Omega_T)$,*

$$\|v\|^2_{2,\Omega_T} \leq \gamma \left| |v| > 0 \right|^{\frac{2}{N+2}} \|v\|^2_{V^2(\Omega_T)} .$$

Lastly we need a measure theoretical lemma that will allow us to cluster the positivity of the solution.

Lemma 2.6 (Measure Theoretical Lemma) *Let $u \in W^{1,1}(K_\rho)$ satisfy*

$$\|u\|_{W^{1,1}(K_\rho)} \leq \gamma\rho^{N-1} \tag{2.7}$$

and $|[u > 1]| \geq \sigma|K_\rho|$ for some $\gamma > 0$ and $\sigma \in (0,1)$. Then for every $\delta \in (0,1)$ and $0 < \lambda < 1$ there exist $x_0 \in K_\rho$ and $\eta = \eta(\alpha, \delta, \gamma, \lambda, N) \in (0,1)$ such that

$$|[u > \lambda] \cap K_{\eta\rho}(x_0)| > (1-\delta)|K_{\eta\rho}(x_0)|. \tag{2.8}$$

This Lemma was established in [12] for $\|u\|_{W^{1,1}(K_\rho)}$ and $p > 1$. It was proved in this form in [14] and was extended to the case of BV spaces in [37].

2.3 Critical Mass Lemma and Shrinking Lemma

The classical De Giorgi approach is based on two lemmata. The first one is named by Caffarelli *Critical Mass Lemma* while the second one is named by DiBenedetto *Shrinking Lemma*. These two lemmata were both proved in [5]. Without loss of generality, we can assume $0 < u < 1$, as the general case can be recovered multiplying the solution u for a constant and adding a second, suitable constant. We also work in unitary cylinders. All we have written here, holds, with the same constants, for parabolic cylinders $Q(R^2, R)$. If one works with cylinders $Q(R^2, S)$ the constants will depend only upon the space dimension N, the ratio S/R and the constants appearing in energy and logarithmic estimates.

Lemma 2.7 (Critical Mass Lemma) *Assume that*

$$u \in \mathrm{L}^2((0,T); W^{1,2}(Q(1,1))) \cap C((0,T); L^2(Q(1,1))) \cap L^\infty(Q(1,1))$$

satisfies the energy estimates (2.4) and $0 < u < 1$. Then there exists an absolute critical value $0 < \nu_0 < 1$, such that if η is any number with $0 < \eta < 1$ and if the measure where $u < \eta$ is smaller than ν_0 than the function u is always greater than $\eta/2$ in the half cube $Q((\frac{1}{2})^2, \frac{1}{2})$.

Proof Let

$$R_n = \frac{1}{2} + \frac{1}{2^{n+1}}, \quad n = 0, 1, \ldots,$$

and build the family of nested and shrinking cylinders $Q(R_n^2, R_n)$. Consider piecewise smooth cutoff functions $0 < \zeta_n \leq 1$, defined in these cylinders, and satisfying the following set of assumptions

$$\zeta_n = 1 \text{ in } Q\left(R_{n+1}^2, R_{n+1}\right), \quad \zeta_n = 0 \text{ on } \partial_p Q\left(R_n^2, R_n\right),$$

$$|\nabla \zeta_n| \leq 2^{n+1}, \qquad 0 \leq \partial_t \zeta_n \leq 2^{2(n+1)},$$

where with ∂_p we denote the parabolic boundary of a set. Write the energy inequality (2.4) for the functions $(u - k_n)_-$, with

$$k_n = \frac{\eta}{2} + \frac{\eta}{2^{n+1}}, \quad n = 0, 1, \ldots,$$

in the cylinders $Q\left(R_n^2, R_n\right)$ and with $\zeta = \zeta_n$, to get:

$$\begin{aligned}
&\sup_{-R_n^2 < t < 0} \int_{K_{R_n} \times \{t\}} (u - k_n)_-^2 \zeta_n^2 \, dx + \int_{-R_n^2}^0 \int_{K_{R_n}} |\nabla (u - k_n)_- \zeta_n|^2 \, dx \, dt \\
&\leq C \int_{-R_n^2}^0 \int_{K_{R_n}} (u - k_n)_-^2 (|\nabla \zeta_n|^2 + \partial_t \zeta_n) \, dx \, dt \\
&\leq C\, 2^{2(n+1)} \int_{-R_n^2}^0 \int_{K_{R_n}} (u - k_n)_-^2 \, dx \, dt.
\end{aligned} \tag{2.9}$$

Next, observing that

$$(u - k_n)_- = \frac{\eta}{2} + \frac{\eta}{2^{n+1}} \leq \eta,$$

we obtain from (2.9)

$$\begin{aligned}
&\sup_{-R_n^2 < t < 0} \int_{K_{R_n} \times \{t\}} (u - k_n)_-^2 \zeta_n^2 \, dx + \int_{-R_n^2}^0 \int_{K_{R_n}} |\nabla (u - k_n)_- \zeta_n|^2 \, dx \, dt \\
&\leq C\, 2^{2(n+1)} (\eta)^2 \int_{-R_n^2}^0 \int_{K_{R_n}} \chi_{\{(u - k_n)_- > 0\}} \, dx \, dt\,.
\end{aligned} \tag{2.10}$$

Define for each n,

$$A_n = \int_{-R_n^2}^{0} \int_{K_{R_n}} \chi_{\{(u-k_n)_- > 0\}} \, \mathrm{d}x \, \mathrm{d}t$$

and observe that

$$\begin{aligned} \frac{1}{2^{2(n+2)}} (\eta)^2 A_{n+1} &\le |k_n - k_{n+1}|^2 A_{n+1} \\ &\le \|(u-k_n)_-\|^2_{2, Q(R^2_{n+1}, R_{n+1})} \\ \le \|(u-k_n)_- \zeta_n\|^2_{2, Q(R_n^2, R_n)} &\le C \, \|(u-k_n)_- \zeta_n\|^2_{V_0^2(Q(R_n^2, R_n))} \, A_n^{\frac{2}{N+2}} \\ &\le C \, 2^{2n} (\eta)^2 A_n^{1+\frac{2}{N+2}} . \end{aligned} \tag{2.11}$$

The first three of these inequalities follow from the definition of A_n and $k_{n+1} < k_n$. The fourth inequality is a consequence of Theorem 2.5 and the last is a consequence of (2.10). Next, define the numbers

$$X_n = \frac{A_n}{|Q(R_n^2, R_n)|} ,$$

divide (2.11) by $|Q(R^2_{n+1}, R_{n+1})|$, and obtain the recursive relation

$$X_{n+1} \le C \, 4^{2n} \, X_n^{1+\frac{2}{N+2}} ,$$

for a constant C depending only upon N. By Lemma 2.4 on fast geometric convergence, if

$$X_0 \le C^{-\frac{N+2}{2}} 4^{-\frac{(N+2)^2}{2}} \equiv \nu_0 \tag{2.12}$$

then

$$X_n \longrightarrow 0 . \tag{2.13}$$

Therefore,

$$\left| \left\{ (x,t) \in Q\left((\tfrac{1}{2})^2, \tfrac{1}{2} \right) \, : \, u(x,t) \le \frac{\eta}{2} \right\} \right| = 0.$$

□

Lemma 2.8 (Shrinking Lemma) *Assume that*

$$u \in \mathrm{L}^2((0,T); W^{1,2}(Q(1,1))) \cap C((0,T); L^2(Q(1,1))) \cap L^\infty(Q(1,1))$$

satisfies the energy estimates (2.4) *and* $0 < u < 1$*. There exist* $0 < \eta < 1$ *and* $0 < \varepsilon < 1$ *such that, if for any time* $t \in (-1,0)$*, the measure of the subset of* K_1 *where* $u(x,t) > \eta$ *is greater than* $\varepsilon(1/2)^N$*, then, for any* $n \in \mathbf{N}$ *in the cylinder* $Q(\frac{1}{2},\frac{1}{2})$

$$|u \le \frac{\eta}{2^n}| \le \frac{C}{\sqrt{n}}, \tag{2.14}$$

where C depends only upon N, ε *and the constant of the energy estimates.*

Proof Take a piecewise smooth cutoff function $0 < \zeta \le 1$, defined in $Q\,(1,1)$, and such that

$$\zeta = 1 \text{ in } Q(\tfrac{1}{2},\tfrac{1}{2}), \quad \zeta = 0 \text{ in } \partial_p Q\,(1,1)\,,$$

$$|\nabla\zeta| \le 2, \qquad 0 \le \partial_t \zeta \le 2.$$

By the energy estimates

$$\int_{-\frac{1}{2}}^{0}\int_{K_{\frac{1}{2}}} |\nabla(u-k)_-|^2\,\mathrm{d}x\,\mathrm{d}t \le C\int_{-1}^{0}\int_{K_1}(u-k)_-^2\,\mathrm{d}x\,\mathrm{d}t$$

$$\le C\,(k)^2\,|Q\,(1,1)|\,. \tag{2.15}$$

Define $B_s \equiv \{(x,t) \in Q(\frac{1}{2},\frac{1}{2}) : u(x,t) \le \frac{\eta}{2^s}\}$. By the previous inequality, we get

$$\int\int_{B_s} |\nabla u|^2\,\mathrm{d}x\,\mathrm{d}t \le C\left(\frac{\eta}{2^s}\right)^2 |Q\,(1,1)|\ . \tag{2.16}$$

Next, apply Lemma 2.3 to the function $u(\cdot,t)$, for all $-1 \le t \le 0$, and with

$$k = \frac{\eta}{2^s}\,, \quad l = \frac{\eta}{2^{s+1}}\,, \quad l-k = \frac{\eta}{2^{s+1}}\,.$$

By our assumptions

$$\left|\left\{x \in K_1 \ :\ u(x,t) > \frac{\eta}{2^s}\right\}\right| \ge \varepsilon,$$

we have

$$\frac{\eta}{2^{s+1}}\,|B_{s+1}(t)| \le \frac{4C}{\varepsilon^2}\int_{B_s(t)\setminus B_{s+1}(t)} |\nabla u|\,\mathrm{d}x\,\mathrm{d}t\ ,$$

for $t \in (-1, 0)$. Integrating over this interval, we conclude that

$$\frac{\eta}{2^{s+1}}\,|B_{s+1}| \le \frac{C}{\varepsilon^2}\int\!\!\int_{B_s\setminus B_{s+1}} |\nabla u|\,\mathrm{d}x\,\mathrm{d}t \le \frac{C}{\varepsilon^2}\left(\int\!\!\int_{B_s} |\nabla u|^2\,\mathrm{d}x\,\mathrm{d}t\right)^{\frac{1}{2}} |B_s \setminus B_{s+1}|^{\frac{1}{2}}$$

$$\le \frac{C}{\varepsilon^2}\left(\frac{\eta}{2^s}\right)|Q\,(1,1)|^{\frac{1}{2}}\ |B_s \setminus B_{s+1}|^{\frac{1}{2}}\ ,$$

using also (2.16). Squaring and dividing through by $\left(\frac{\eta}{2^{s+1}}\right)^2$, we obtain

$$|B_{s+1}|^2 \le C\,(\varepsilon)^4\ |Q\,(1,1)|\ |B_s \setminus B_{s+1}|\ .$$

Since these inequalities are valid for $0 \le s \le s_0$, we add them to get

$$s_0\ \left|B_{s_0}\right|^2 \le C\,(\varepsilon)^4\,|Q\,(1,1)|^2\ ,$$

that is,

$$\left|B_{s_0}\right| \le \frac{C}{(s_0)^{\frac{1}{2}}}\,(\varepsilon)^2\ |Q\,(1,1)|$$

and the last inequality implies (2.14). □

2.4 *Expansion of Positivity Lemmata*

This approach was developed by DiBenedetto and his school in the last years and it is a powerful tool that sometimes works also where other approaches are not more useful. Also in this subsection we can assume $0 < u < 1$.

Lemma 2.9 (Clustering the Positivity) *Assume that*

$$u \in \mathrm{L}^2((0,T);\,W^{1,2}(Q(1,1))) \cap C((0,T);\,L^2(Q(1,1))) \cap L^\infty(Q(1,1))$$

satisfies the energy estimates (2.4) and $0 < u < 1$. Assume that there exists $\eta > 0$ such that $|Q(1/2,1/2)) \cap \{(x,t) : u(x,t) > 1/2\}| \ge \eta$. Then for any $0 < \alpha < 1/2$ and $0 < \nu < 1$ there is a number $\varepsilon_0 > 0$ that can be quantitatively determined and *that depends only upon N, α and ν, such that there exist two values $x_0 \in K_{1/2}$ and $t_0 \in (-1/2, 0)$ such that*

$$|\{x \in K_\varepsilon(x_0) : u(x, t_0) \geq \alpha\}| \geq (1 - \nu)|K_\varepsilon|. \tag{2.17}$$

Proof Take a piecewise smooth cutoff function $0 < \zeta \leq 1$, defined in $Q(1, 1)$, and such that

$$\zeta = 1 \text{ in } Q(\tfrac{1}{2}, \tfrac{1}{2}), \quad \zeta = 0 \text{ in } \partial_p Q(1, 1),$$

$$|\nabla\zeta| \leq 2, \qquad 0 \leq \partial_t \zeta \leq 2.$$

By the energy estimates

$$\int_{-\frac{1}{2}}^0 \int_{K_{\frac{1}{2}}} \left|\nabla u^2\right| \, dx\, dt \leq C \int_{-1}^0 \int_{K_1} u^2 \, dx\, dt \leq C. \tag{2.18}$$

Define the set

$$A = \{t \in (-1/2, 0) : \int_{K_{1/2}} \left|\nabla u^2(t)\right| \, dx\, dt \leq \frac{16\, C}{\eta}\}$$

where C is the constant appearing in (2.18). Clearly $|A| \geq 1/2 - \frac{\eta}{16}.$ Moreover, define the set

$$B = \{t \in (-1/2, 0) : |x \in K_{1/2} : u(x, t) > 1/2| > \eta/2.$$

Clearly $|B| \geq \eta/2$. Hence $|A \cap B| \geq \eta/4 > 0$.

Now, choose a time level $t \in A \cap B$, call $v = 2u$ and note that

$$\|v(t)\|_{W^{1,1}(K_{1/2})} \leq \|v(t)\|_{W^{1,p}(K_{1/2})} \frac{1}{2^{\frac{(p-1)N}{p}}} \leq 2\left(\frac{16\, C}{\eta}\right)^{\frac{1}{p}} \frac{1}{2^{\frac{(p-1)N}{p}}}$$

and

$$|[v(t) > 1]| \geq \eta/2|K_{1/2}|.$$

Now (2.17) comes from Lemma 2.7 in a straightforward way. □

Lemma 2.10 (Positivity Everywhere) *Assume that*

$$u \in L^2((0, T); W^{1,2}(Q(1, 1))) \cap C((0, T); L^2(Q(1, 1))) \cap L^\infty(Q(1, 1))$$

satisfies the energy and logarithmic estimates and $u \geq 0$. Assume that $|u(x,-1) > 1| \geq (1-\nu_0/2)$ where ν_0 is the constant appearing in the Critical Mass Lemma. Then there exists a number $\varepsilon > 0$, depending only upon N and the constants appearing in the energy and logarithmic estimates, such that $u \geq \varepsilon$ everywhere in $Q(1/2, 1/2)$.

Proof The proof consists in using the logarithmic inequalities (2.5) applied to the function $(u-1)_-$ in the cylinder $K_1 \times (-1, t^*)$, with $-1 < t^* < 0$ and

$$k = 1 \quad \text{and} \quad c = \frac{1}{2^n},$$

where $n \in \mathbb{N}$ will be chosen later.

Define

$$H^-_{u,1} = \operatorname*{ess\,sup}_{K_1 \times (0,t^*)} \left|(u-1)_-\right| . \tag{2.19}$$

If $H^-_{u,1} \leq \frac{1}{2}$ the statement of the lemma is trivially verified. Assuming $H^-_{u,1} > \frac{1}{2}$, the logarithmic function Ψ^- is defined in the whole $K_1 \times (0, t^*)$, and it is given by

$$\Psi^- = \psi^-_{\left\{H^-_{u,1}, \frac{1}{2^{n+1}}\right\}}(u) = \begin{cases} \ln \left\{ \dfrac{H^-_{u,1}}{H^-_{u,1} - 1 + u + \frac{1}{2^{n+1}}} \right\} & \text{if } u < 1 - \frac{1}{2^{n+1}} \\[2ex] 0 & \text{if } u \geq 1 - \frac{1}{2^{n+1}} \end{cases}$$

By straightforward calculation one gets

$$\Psi^- \leq (n+1)\ln 2.$$

Choosing a piecewise smooth cutoff function $0 < \zeta(x) \leq 1$, defined in K_1 and such that, for some $\sigma \in (0, 1)$,

$$\zeta = 1 \text{ in } K_{(1-\sigma)} \quad \text{and} \quad |\nabla \zeta| \leq (\sigma)^{-1} ,$$

inequality (2.5) becomes

$$\sup_{-1<t<t^*} \int_{K_1 \times \{t\}} \left[\psi^-(u)\right]^2 \zeta^2 \, dx \leq \int_{K_1 \times \{-1\}} \left[\psi^-(u)\right]^2 \zeta^2 \, dx$$
$$+ C \int_{-1}^{t^*} \int_{K_1} \psi^-(u) |\nabla \zeta|^2 \, dx \, dt . \tag{2.20}$$

The first integral on the right-hand side can be bounded above taking into account that Ψ^- vanishes on the set $\left\{x \in K_1 \;:\; u(x,\cdot) \geq 1 - \frac{1}{2^{n+1}}\right\}$. This gives

$$\int_{K_1 \times \{-1\}} \left[\psi^-(u)\right]^2 \,\mathrm{d}x \leq n^2 (\ln 2)^2 \,\nu_0/2\, |K_1| \;.$$

The second integral can be estimated in this way:

$$C \int_{-1}^{t^*} \int_{K_1} \psi^+(u) \;\; |\nabla \zeta|^2 \,\mathrm{d}x\,\mathrm{d}t$$
$$\leq C\,(n+1)(\ln 2)\,\sigma^{-2}\,(t^*+1)\,|K_1|\,.$$

The left-hand side is estimated below by integrating over the smaller set

$$S = \left\{x \in K_{(1-\sigma)} \;:\; u(x,t) > \frac{1}{2^{n+1}}\right\} \subset K_1$$

and observing that in S, $\zeta = 1$ and

$$\frac{H_u^-}{H_u^- - 1 + u + \frac{1}{2^{n+1}}} \geq \frac{1}{\frac{1}{2^n}} = 2^n.$$

Therefore,

$$\left[\psi^-(u)\right]^2 \geq (n)^2 (\ln 2)^2 \,.$$

From this reason

$$\sup_{-1<t<t^*} \int_{K_1 \times \{t\}} \left[\psi^-(u)\right]^2 \,\zeta^2 \,\mathrm{d}x \geq n^2 (\ln 2)^2 \,|S|\,.$$

Combining these three estimates, we arrive at

$$|S| \leq \left(\frac{n+1}{n}\right)^2 \nu_0/2 \;|K_1| + C\,\frac{n+1}{n^2}\,\frac{1}{\sigma^2}(t^*+1)\;|K_1|\,.$$

On the other hand

$$\left|\left\{x \in K_1 \ : \ u(x,t) < \frac{1}{2^{n+1}}\right\}\right|$$

$$\leq \left|\left\{x \in K_{(1-\sigma)} \ : \ u(x,t) < \frac{1}{2^{n+1}}\right\}\right| + \left|K_1 \setminus K_{(1-\sigma)}\right|$$

$$\leq |S| + N\sigma \ |K_1| ;$$

thus

$$\left|\left\{x \in K_1 \ : \ u(x,t) < \frac{1}{2^{n+1}}\right\}\right|$$

$$\leq \left\{\left(\frac{n+1}{n}\right)^2 \nu_0/2 + \frac{C}{n\sigma^2} + N\sigma(t^* + 1)\right\} |K_R| \ ,$$

for all $t \in (-1, t^*)$. Choose σ so small that $N\sigma(t^* + 1) \leq \frac{1}{8}\nu_0$ and n so large that

$$\left(\frac{n+1}{n}\right)^2 \leq \frac{5}{4}$$

and

$$\frac{C}{n\sigma^2} \leq \frac{1}{8}\nu_0.$$

Choose $t^* = 0$ to get that in the unitary cylinder $Q(1, 1)$ the set where $u < \frac{1}{2^{n+1}}$ has a measure smaller than the critical value ν_0. Hence, by the Critical Mass Lemma, $u \geq \frac{1}{2^{n+2}}$ in $Q(1/2, 1/2)$).

If $u(x, -1) > 1$ for any $x \in K_1$ the expansion of positivity is true for any time. This result was proved, for the first time, in [12]. □

Lemma 2.11 (Expansion of Positivity in Time) *Let* $\Omega_\infty \equiv K_1 \times (-1, \infty)$. *Assume that*

$$u \in L^2((0, T); W^{1,2}(\Omega_\infty)) \cap C((0, T); L^2(\Omega_\infty)) \cap L^\infty(\Omega_\infty)$$

satisfies the energy and logarithmic estimates and $u \geq 0$. *Assume that* $u(x, -1) > 1$. *Then*

$$u(x,t) \geq Ce^{-\alpha_0(t+1/2)}$$

for any $(x, t) \in K_{1/2} \times (-1/2, \infty)$ *where the constants* C *and* α_0 *depend only upon* N *and the constant appearing in the energy and logarithmic estimates.*

Proof Let us repeat the argument of the previous lemma choosing $\sigma = 1/2$. Noting that the contribution of the first integral on the right hand side vanishes, we get

$$|S| \le C\,\frac{n+1}{n^2}\,4(t^*+1)\;|K_1|$$

Choose n so large that

$$C\,\frac{n+1}{n^2}\,4(t^*+1)2^N < \nu_0.$$

This implies that one can choose $n = \alpha_0(t^*+1)$ with α_0 a suitable positive constant. This fact implies the statement. □

Combining the previous lemmata one gets the following result:

Theorem 2.12 (Time Expansion) *Let* $\Omega_\infty \equiv K_1 \times (-1, \infty)$. *Assume that*

$$u \in \mathrm{L}^2((0,T); W^{1,2}(\Omega_\infty)) \cap C((0,T); L^2(\Omega_\infty)) \cap L^\infty(\Omega_\infty)$$

satisfies the energy and logarithmic estimates and $0 \le u \le 1$. *Assume that there exists* $\eta > 0$ *such that* $|Q(1/2, 1/2)) \cap \{(x,t) : u(x,t) > 1/2\}| \ge \eta$. *Then there exist* $x_0 \in K_{1/2}$, $t_0 \in (0,1)$ *and positive constants* r, α, C *(depending only upon* N *and the constants appearing in the energy and logarithmic estimates) such that*

$$u(x,t) \ge Ce^{-\alpha t} \tag{2.21}$$

for any $(x, t) \in K_r(x_0) \times (t_0, \infty)$.

3 Linear Equations: Regularity Results and Harnack Estimates

3.1 Regularity of the Solutions

To prove the regularity of the solution, here we will follow an approach proposed in [21] that has the advantage to be very close to the one we will use to prove the Harnack estimates.

Theorem 3.1 (Hölder) *Let* $\Omega_\infty \equiv K_1 \times (0,T)$ *where* T *is a positive number. Assume that* $u \in \mathrm{L}^2((0,T); W^{1,2}(\Omega_T)) \cap C((0,T); L^2(\Omega_T))$ *satisfies the energy and logarithmic estimates. Then there exists* $\alpha > 0$ *such that, for any compact* $K \subset\subset \Omega_T$ *we have that* $u \in C^{0,\alpha}(K)$. *Moreover the constant of Hölder continuity* *depends only upon N, the distance between K and $\partial_p \Omega_T$ and the constants appearing in the energy and logarithmic estimates.*

Proof Let $(x_0, t_0) \in K$. Define $Q_n = (x_0, t_0) + Q(4^{-2n}, 4^{-n})$ and $\omega_n \equiv \text{esssup}_{Q_n} u$. Define $4^n = \rho^n$ Assume that there is a constant $\eta \in (0, 1)$, that can be determined a priori only in terms of the data, such that for every $(x_0, t_0) \in K$ and every $n \in \mathbb{N}$,

$$\omega_{n+1} \leq (1 - \eta)\omega_n. \tag{3.1}$$

This inequality implies that u is locally Hölder continuous in Ω_T. Actually, by iteration,

$$\omega_n \leq (1 - \eta)^n \omega_0 , \qquad \forall n \in \mathbb{N}. \tag{3.2}$$

Define $(1 - \eta) = \delta^\alpha$ and $\alpha = \frac{\ln(1-\eta)}{\ln \delta} \in (0, 1)$. Therefore,

$$\omega_n \leq \omega_0 \left(\frac{\rho_n}{\rho_0}\right)^\alpha , \qquad \forall n \in \mathbb{N}. \tag{3.3}$$

Since $(x_0, t_0) \in \Omega_T$ is arbitrary, we conclude that u is locally Hölder continuous in K with exponent α. To end the proof, we have to show that (3.1) holds. □

Let us prove (3.1). Without loss of generality we can assume $(x_0, t_0) = (0, 0)$, $\omega = 1$, and $0 < u < 1$ Via a homothetical transformation we can assume to work in the cylinder $Q(16, 4)$ and prove the reduction of oscillation in $Q(1, 1)$.

We follow several steps:

First Step: Localization of Positivity in a Subcylynder Consider the cylinder $(0, -15) + Q(1, 1)$ that lies at the bottom of $Q(16, 4)$. We may assume that in this cylinder the measure where $u \geq 1/2$ is greater of $1/2$ (otherwise one works with the function $v = 1 - u$). By Theorem 2.12 there exists $x_0 \in K_{1/2}$, $t_0 \in (-15, -14)$ and positive constants r, α, C (depending only upon N and the constants appearing in the energy and logarithmic estimates) such that

$$u(x, t) \geq Ce^{-\alpha(t_0 - t)}$$

for any $(x, t) \in K_r(x_0) \times (t_0, 0)$. In particular, $u(x, t) \geq Ce^{-15}$ for any $(x, t) \in K_r(x_0) \times (-9, 0)$

Second Step: The Shrinking Lemma Apply the Shrinking Lemma to the cylinder $Q(9, 3)$ to get that there exists a positive constant $\eta > 0$, such that in $Q(4, 2)$ the measure where $u > 2\eta$ is greater than $(1 - \nu)|Q(4, 2)|$.

Third Step: The Critical Mass Lemma Apply the Critical Mass Lemma to the cylinder $Q(4,2)$ to get that there $u > \eta$ in $Q(1,1)$. This implies that the oscillation of u in $Q(1,1)$ is smaller than $1-\eta$.

3.2 Harnack Estimates

The first parabolic version of the Harnack inequality was due to Hadamard (see [23]) and Pini (see [33]). The proof is based on local representations by means of heat potentials. A breakthrough in the theory is due to Moser, who in his celebrated paper [30] proved that Harnack inequality continues to hold for nonnegative weak solution of the type

$$\begin{cases} u \in C_{\text{loc}}\left(0,T;L^2_{\text{loc}}(\Omega)\right) \cap L^2_{\text{loc}}\left(0,T;W^{1,2}_{\text{loc}}(\Omega)\right) \\ u_t - \sum_{i,j=1}^{N} D_i(a_{ij}(x,t)D_j u) = 0 \qquad \text{in } \Omega_T, \end{cases} \tag{3.4}$$

where $a_{ij} \in L^\infty(\Omega_T)$ and satisfy the ellipticity condition

$$\sum_{i,j=1}^{N} a_{ij}\xi_i\xi_j \geq \nu|\xi|^2, \qquad \forall \xi \in \mathbb{R}^N \tag{3.5}$$

with ν a positive constant. The result of Moser can be extended (see [1, 39] and [38]) to nonnegative weak solutions of the quasilinear parabolic equation

$$u_t - \operatorname{div} \mathbf{a}(x,t,u,\nabla u) + b(x,t,u,\nabla u) = 0 \qquad \text{in } \Omega_T, \tag{3.6}$$

The original proof by Moser was based on the John-Nirenberg Lemma [24]. Moser himself published a new proof of Harnack inequality in 1971 (see [31]), with the expressed purpose to avoid the use of the parabolic John-Nirenberg Lemma. He got this result using estimates on the logarithm of the solution and a tricky measure lemma based on a result by Bombieri [3, 4]. A few years later Fabes and Garofalo (see [18]) came back to Moser's Main Lemma and gave a simplified proof, using Calderon's proof of the original John-Nirenberg lemma (see also[19]).

Here we sketch a more direct proof that holds also in the setting of the De Giorgi classes (the first proof of this fact is in [8]). Even though not so explicitly stated, a similar method is already present in the work by Krylov and Safonov [25].

Theorem 3.2 (Harnack) *Let u be a nonnegative function in Ω_T. Assume that*

$$u \in \mathrm{L}^2((0,T);W^{1,2}(\Omega_T)) \cap C((0,T);L^2(\Omega_T))$$

satisfies the energy and logarithmic estimates. Let $(x_0, t_0) \in \Omega_T$ *and assume that the cylinder* $(x_0, t_0) + Q_{2\rho} \subset \Omega_T$*, where* $Q_\rho \equiv B_\rho \times (-\rho^2, 0)$*. Then there exists a constant* γ*, depending only upon* N *and the constants appearing in the energy and logarithmic estimates, such that*

$$u(x_0, t_0) \geq \gamma \sup_{B_\rho(x_0)} u(x, t_0 - \rho^2) \,. \tag{3.7}$$

Proof Also in this case we follow several steps:

First Step: Renormalization of the Solution Without loss of generality we can assume that $(x_0, t_0) = (0, 0)$, $u(0, 0) = 1$ and $\rho = 2$.

Second Step: Determination of the Largest Value of u in $Q(1, 1)$ For the sake of simplicity, assume that the function u is continuous. Build the family of nested boxes $Q_\tau \equiv B_\tau \times (-\tau^2, 0]$. Define the numbers $M_\tau = \sup_{Q_\tau} u$ and $N_\tau = (1 - \tau)^{-\beta}$ where $\beta > 1$ will be chosen later. Let $0 \leq \tau_0 < 1$ be the largest root of the equation $M_\tau = N_\tau$. Such a root is well defined, since $M_0 = N_0$, and as $\tau \to 1^-$ M_τ remains bounded, whereas N_τ blows up. By construction, $\sup_{Q_\tau} v \leq N_\tau$ for all $\tau > \tau_0$. Moreover, from the continuity of v in Q, there exists at least a point $(x_1, t_1) \in N_{\tau_0}$ where $u(x_1, t_1) = (1 - \tau_0)^{-\beta}$.

Third Step: Lower Bound on u at the Same Time-Level t_1 Consider the cylinder $\tilde{Q}_\tau \equiv (x_1, t_1) + Q((\frac{(1-\tau_0)}{2})^2, \frac{(1-\tau_0)}{2})$: in such a cylinder we have that $u \leq 2^\beta (1 - \tau_0)^{-\beta}$. The Critical Mass Lemma yields that

$$\left| (x, t) \in \tilde{Q}_\tau : u(x, t) \geq \frac{(1 - \tau_0)^{-\beta}}{4} \right| \geq \nu |\tilde{Q}_\tau|.$$

because, otherwise, one could apply the Critical Mass Lemma and conclude that $u(x_1, t_1) \leq \frac{(1-\tau_0)^{-\beta}}{2}$, which is a contradiction. Hence, by arguing as in the clustering of positivity lemma, there is a point $(x_2, t_2) \in \tilde{Q}_\tau$ such that we can determine a small ball of radius r_0 *qualitatively* determined. about (x_2, t_2) where $u \geq \frac{(1-\tau_0)^{-\beta}}{4}$.

Fourth Step: Time Expansion of the Positivity Set By Theorem 2.12 there exists α_1, depending only upon N, β, τ_0 and the constants appearing in the energy and logarithmic estimates, and a t_2 such that

$$u(x, t) \geq C e^{-\alpha_1 (t_2 - t)}$$

for any $(x, t) \in K_r(x_2 \times (t_2, \infty)$.

Fifth Step: Space Expansion of the Positivity Set From this point till the end, the proof is similar to the one given in the steps two and three of the regularity. The only difference comes from the necessity to do a suitable choice of the parameter β. For more details see the monograph [10].

□

4 The p-Laplacean Equation

4.1 Statement of the Results: Regularity and Harnack Estimates

In this section we consider bounded, weak solutions of the p-Laplacean equation

$$u_t = \sum_{i=1}^{N} \operatorname{div}(|D_i u|^{p-2} D_i u) \quad p > 2 \tag{4.1}$$

The regularity for solutions of the p-Laplacean with bounded and measurable coefficients was proved in [7]. For the regularity of the prototype equation see [11, 40]. In this section we state the regularity results for solution of the p-Laplacean following the proof given in [21]. Note that the original proof given by DiBenedetto is more general of the one sketched here, because it works also for functions that belong to a suitable De Giorgi class.

Theorem 4.1 (Hölder Continuity) *Let $\Omega_T \equiv K_1 \times (0, T)$ where T is a positive number. Assume that*

$$u \in \mathrm{L}^p((0, T); W^{1,p}(\Omega_T)) \cap C((0, T); L^2(\Omega_T))$$

is a weak solution of (4.1). Then there exists $\alpha > 0$ such that, for any compact $K \subset\subset \Omega_T$ we have that $u \in C^{0,\alpha}(K)$. Moreover, the constant of Hölder continuity depends only upon the data.

In Moser's approach the main feature that makes the method work is the homogeneity of the time and space terms of the equation; in fact Trudinger [39, Section 5] shows that things run in the same way in the proof of Harnack inequality for doubly nonlinear equations of the type

$$(u^{p-1})_t - \operatorname{div}(|\nabla u|^{p-2} \nabla u) = 0 \qquad \text{in } \Omega_T, \tag{4.2}$$

which is p-homogeneous, exactly as (3.4) is 2-homogeneous.

On the other hand, coming back to equation (4.1), quite surprisingly Moser's method does not work when $p \neq 2$ and this is not simply a matter of technique.

In the case of the p-Laplacean a first result is due to DiBenedetto in [9]. He proved the Harnack estimates for the prototype equations by using the Barenblatt solutions, i.e. the explicit solutions for the p-Laplacean introduced by Barenblatt in [2]. The general result was proved in [15]

Theorem 4.2 (Harnack) *Let u be a nonnegative weak solution of (4.1) and let $p > 2$. Fix $(x_0, t_0) \in \Omega_T$ and assume that $u(x_0, t_0) > 0$. There exist constants $\gamma > 1$ and $C > 0$, depending only upon N and p, such that*

$$u(x_0, t_0) \leq \gamma \inf_{B_\rho(x_0)} u(\cdot, t_0 + \theta)\,, \tag{4.3}$$

where

$$\theta \equiv \frac{C\rho^p}{[u(x_0, t_0)]^{p-2}}, \tag{4.4}$$

provided that the cylinder $(x_0, t_0) + B_{4\rho} \times (-4\theta, 4\theta)$ is contained in Ω_T.

Here we do not give the proof of the previous two theorems. We only focus on the main differences between the linear case and the degenerate one.

4.2 *Differences between the Linear Case and the Degenerate One*

Intrinsic Geometry

In the degenerate cases the estimates are not homogeneous, in the sense that they involve integral norms corresponding to different powers, namely the powers 2 and p. The key idea is then to look at the equation in its own geometry, i.e., in a geometry dictated by its intrinsic structure. This amounts to rescale the standard parabolic cylinders by a factor that depends on the oscillation of the solution. This procedure, which can be called accommodation of the degeneracy, allows one to recover the homogeneity in the energy estimates written over these rescaled cylinders. We can heuristically say that the equation behaves in its own geometry like the heat equation. Such kind of geometry is called *intrinsic*, because it is intrinsic to the solution itself. For more details see the monograph [41]. Let us consider the energy estimates, which in the degenerate setting become:
if u is a local weak solution of (4.1) then there exists a constant $C \equiv C(p) > 0$ such that for every cylinder $Q(\tau, \rho) \subset \Omega_T$,

$$\begin{aligned}
&\sup_{-\tau<t<0} \int_{K_\rho \times \{t\}} (u-k)_\pm^2\, \zeta^p \,dx + \int_{-\tau}^0 \int_{K_\rho} |\nabla (u-k)_\pm \zeta|^p \,dx\,dt \\
&\leq \int_{K_\rho \times \{-\tau\}} (u-k)_\pm^2\, \zeta^p \,dx + C \int_{-\tau}^0 \int_{K_\rho} (u-k)_\pm^p\, |\nabla \zeta|^p \,dx\,dt \\
&\quad + p \int_{-\tau}^0 \int_{K_\rho} (u-k)_\pm^2\, \zeta^{p-1}\, \partial_t \zeta \,dx\,dt\,.
\end{aligned} \tag{4.5}$$

Now, roughly speaking, the term

$$\int_{-\tau}^{0}\int_{K_\rho}(u-k)_\pm^p\,|\nabla\zeta|^p\,\mathrm{d}x\,\mathrm{d}t$$

goes as $|A_k|\omega^p\rho^{-p}$ where A_k is the set in $Q(\tau,\rho)$ where $u_>{}^<k$ and ω is the oscillation of u.

The term

$$\int_{-\tau}^{0}\int_{K_\rho}(u-k)_\pm^2\,\zeta^{p-1}\,\partial_t\zeta\,\mathrm{d}x\,\mathrm{d}tt$$

goes as $|A_k|\omega^2\tau^{-1}$. As proved in [36], if the two terms are not homogeneous, the functions belonging to such a class could be discontinuous. Therefore, the only way to make homogeneous the two terms is to take $\tau = R^p\omega^{2-p}$. Hence, the more the oscillations go to zero, the more the cylinders become longer in the time direction. This is the reason why we are compelled to work in the intrinsic geometry.

The p-Geometry

All the statements must be reformulated in terms of the fact that the gradient of the solution is also in L^p and not only in L^2. For instance, in this setting, the theorem about the time expansion becomes

Theorem 4.3 (Time Expansion for the *p*-Laplacean) *Let $\Omega_\infty \equiv K_1\times(-1,\infty)$. Assume that*

$$u\in \mathrm{L}^p((0,T);W^{1,p}(\Omega_\infty))\cap C((0,T);L^2(\Omega_\infty))\cap L^\infty(\Omega_\infty)$$

satisfies the p-energy and the p-logarithmic estimates, and $0\le u\le 1$. Assume that there exists $\eta>0$ such that $|Q((1/2)^p,1/2))\cap\{(x,t):u(x,t)>1/2\}|\ge\eta$. Then there exist $x_0\in K_{1/2}$, $t_0\in(0,1)$ and positive constants r,α,C (depending only upon N and the constants appearing in the energy and logarithmic estimates), such that

$$u(x,t)\ge C(t+1)^{-\frac{1}{p-2}} \tag{4.6}$$

for any $(x,t)\in K_r(x_0)\times(t_0,\infty)$.

A Tricky Change of Variables

The problem in applying the approach described for the linear case, is that when one applies the Shrinking Lemma, one gets the estimates for a smaller value of the solution and in order to apply the Critical Mass Lemma, one needs a different geometry. In the case of the proof of the Hölder continuity, DiBenedetto was able to solve this problem thanks to an approach based on an alternative argument.

This method does not work to prove the Harnack estimates for a general equation. The DiBenedetto's proof in [9] relies only on the fact that explicit solutions are known. The key point to overcome this difficulty is a tricky change of variables that was introduced in [20] and improved and generalized in [15]. This change of variables can be applied also to get spacial expansion of positivity (see, for instance [17, 27, 28]).

Let us start from the estimates of expansion of positivity:

$$u(x,t) \geq C(t+1)^{-\frac{1}{p-2}}$$

and apply the Shrinking Lemma in a cylinder of length t_0. We have that $u \geq C(t_0+1)^{-\frac{1}{p-2}}$ in a sizeable portion of space for any time t belonging to $(0, t_0)$. Then by the Shrinking Lemma $u \geq \eta(t_0+1)^{-\frac{1}{p-2}}$. In order to apply the Critical Mass Lemma (as the length of the cylinder in the intrinsic geometry depends on how small is the function) we need to apply the Shrinking Lemma in a longer cylinder, we say, till a time $t_1 > t_0$. But there the estimates on u is deteriorating because we now have that $u \geq C(t_1+1)^{-\frac{1}{p-2}}$. Therefore, applying once again the Shrinking Lemma, we get that $u \geq \eta(t_1+1)^{-\frac{1}{p-2}}$ and, now, the Critical Mass Lemma, to be applied, requires a longer cylinder till the time $t_2 > t_1$, and so on.

The idea to exit this infinite loop is the following. Consider the function $v = (t+1)^{\frac{1}{p-2}} u$ and note that

$$v(x,t) \geq C$$

for any $t > 0$. The function v satisfies the equation

$$t v_t = \sum_{i=1}^{N} \operatorname{div}(|D_i v|^{p-2} D_i v) - \frac{1}{p-2} v$$

that is an Euler equation. If one operates the change of variables $t = e^\tau$ one gets that $w(x,\tau) = v(x, \log t)$ solves the equation

$$w_\tau = \sum_{i=1}^{N} \operatorname{div}(|D_i w|^{p-2} D_i w) - \frac{1}{p-2} w.$$

Now, noting that w is a supersolution of the equation

$$w_\tau = \sum_{i=1}^{N} \operatorname{div}(|D_i w|^{p-2} D_i w)$$

Fig. 2 Emmanuele with Paolo Marcellini, Ugo Gianazza and Vincenzo Vespri at the Math Department of Vanderbilt University

one can apply the Shrinking Lemma and Critical Mass Lemma to w without any problem, because the estimate on w is not deteriorating in longer cylinders. In this way one obtains both the required reduction of the oscillation for w (and hence for u) and the space expansion of positivity necessary to prove the Harnack estimates (Fig. 2).

References

1. D.G. Aronson, J. Serrin, Local behaviour of solutions of quasilinear parabolic equations. Arch. Rat. Mech. Anal. **25**, 81–123 (1967)
2. G.I. Barenblatt, On some unsteady motions of a liquid or a gas in a porous medium. Prikl. Mat. Mech. **16**, 67–78 (1952)
3. E. Bombieri, *Theory of Minimal Surfaces and a Counterexample to the Bernstein Conjecture in High Dimension*. Mimeographed Notes of Lectures held at Courant Institut (New York University, New York, 1970)
4. E. Bombieri, E. Giusti, Harnack's inequality for elliptic differential equations on minimal surfaces. Inventiones Math. **15**, 24–46 (1972)
5. E. De Giorgi, Sulla differenziabilità e l'analiticità delle estremali degli integrali multipli regolari. Mem. Accad. Sci. Torino. Cl. Sci. Fis. Mat. Nat. **3**(3), 25–43 (1957)
6. E. DiBenedetto, Continuity of weak solutions to certain singular parabolic equations. Ann. Mat. Pura Appl. CXXI (4) **130**(1), 131–176 (1982)
7. E. DiBenedetto, On the local behaviour of solutions of degenerate parabolic equations with measurable coefficients. Ann. Sci. Norm. Sup. **13**, 487–535 (1986)
8. E. DiBenedetto, *Harnack Estimates in Certain Function Classes*. Atti Seminario Matematico e Fisico University Modena XVI, Invited paper for the 70th birthday of Renato Nardini (1988), 173–182
9. E. DiBenedetto, Intrinsic Harnack type inequalities for solutions of certain degenerate parabolic equations. Arch. Rat. Mech. Anal. **100**, 129–147 (1988)
10. E. DiBenedetto, *Degenerate Parabolic Equations* (Springer/Series Universitext, New York, 1993)
11. E. DiBenedetto, A. Friedman, Regularity of solutions of non−linear degenerate parabolic systems. J. Reine Angew. Math. **349**, 83–128 (1984)
12. E. DiBenedetto, V. Vespri, On the singular equation $\beta(u)_t = \Delta u$. Arch. Rational Mech. Anal. **132**, 247–309 (1995)
13. E. DiBenedetto, J. M. Urbano, V, Vespri, Current issues on singular and degenerate evolution equations. *Evolutionary Equations*, vol. I. Handbook Differential Equation (North-Holland, Amsterdam, 2004), pp. 169–286
14. E. DiBenedetto, U. Gianazza, V. Vespri, Local clustering of the non-zero set of functions in $W^{1,1}(E)$. Atti Accad. Naz. Lincei Cl. Sci. Fis. Mat. Natur. Rend. Lincei (9) Mat. Appl. **17**, 223–225 (2006)
15. E. DiBenedetto, U. Gianazza, V. Vespri, Harnack estimates for quasi linear degenerate parabolic differential equations. Acta Math **200**(2), 181–209 (2008)
16. E. DiBenedetto, U. Gianazza, V. Vespri, Harnack's inequality for degenerate and singular parabolic equations, in *Springer Monographs in Mathematics* (Springer, New York, 2012)
17. F.G. Düzgün, P. Marcellini, V. Vespri, An alternative approach to the Hölder continuity of solutions to some elliptic equations. NonLinear Anal. **94**, 133–141 (2014)
18. E.B. Fabes, N. Garofalo, Parabolic B.M.O. and Harnack's inequality. Proc. Am. Math. Soc. **95**, 63–69 (1985)
19. E.B. Fabes, D.W. Stroock, A new proof of Moser's parabolic Harnack inequality using the old ideas of Nash. Arch. Rational Mech. Anal. **96**, 327–338 (1986)
20. U. Gianazza, V. Vespri, A Harnack inequality for a degenerate parabolic equation. J. Evol. Equ. **6**(2), 247–267 (2006)
21. U. Gianazza, M. Surnachev, V. Vespri, On a new proof of Hölder continuity of solutions of p-Laplace type parabolic equations. Adv. Calc. Var. **3**, 263–278 (2010)
22. D. Gilbarg, N. Trudinger, *Elliptic Partial Differential Equations of Second Order* (Springer, New York, 1983)
23. J. Hadamard, Extension à l'équation de la chaleur d'un théorem de A. Harnack. Rend. Circ. Mat. Palermo (2) **3**, 337–346 (1954)

24. F. John, L. Nirenberg, On functions of bounded mean oscillation. Comm. Pure Appl. Math. **14**, 415–426 (1961)
25. N.V. Krylov, M.V. Safonov, A certain property of solutions of parabolic equations with measurable coefficients. Math. USSR Izvestija **16**(1), 151–164 (1981)
26. O.A. Ladyzhenskaja, V.A. Solonnikov, N.N. Ural'ceva, *Linear and quasilinear equations of parabolic type*, in *AMS Translational Mathematical Monographical*, vol. 23 (AMS, Providence RI, 1968)
27. N. Liao, I.I. Skrypnik, V. Vespri, Local regularity for an anisotropic elliptic equation. Calc. Var. PDEs **59**, 1–31 (2020)
28. V. Liskevich, I.I. Skrypnik, Hölder continuity of solutions to an anisotropic elliptic equation. Nonlinear Anal. **71**, 1699–1708 (2009)
29. J. Moser, A new proof of DeGiorgi's theorem concerning the regularity problem for elliptic differential equations. Comm. Pure Appl. Math. **13**, 457–468 (1960)
30. J. Moser, A Harnack inequality for parabolic differential equations. Comm. Pure Appl. Math. **17**, 101–134 (1964)
31. J. Moser, On a pointwise estimate for parabolic differential equations. Comm. Pure Appl. Math. **24**, 727–740 (1971)
32. J. Nash, Continuity of solutions of parabolic and elliptic equations. Am. J. Math. **80**, 931–954 (1958)
33. B. Pini, Sulla soluzione generalizzata di Wiener per il primo problema di valori al contorno nel caso parabolico. Rend. Sem. Math. Padova **23**, 422–434 (1954)
34. J. Schauder, Über lineare elliptische Differentialgleichungen zweiter Ordnung. Math. Z. **38**(1), 257–282 (1934)
35. J. Schauder, Numerische Abschätzungen in elliptischen linearen Differentialgleichungen. Studia Math. Polska Akademia Nauk. Instytut Matematyczny, **5**, 34–42 (1937)
36. S. Spanne, Some function spaces defined using the mean oscillation over cubes. Ann. Scuola Norm. Sup. Pisa **19**(3), 593–608 (1965)
37. A. Telcs, V. Vespri, A Quantitative Lusin theorem for functions in BV, in *Proceedings of a Cortona Conference. Springer. INdAM Series. Geometric Methods in PDEs* (Springer, Berlin, 2015), pp. 81–87
38. N.S. Trudinger, On Harnack type inequalities and their application to quasilinear elliptic partial differential equations. Comm. Pure Appl. Math. **20**, 721–747 (1967)
39. N.S. Trudinger, Pointwise estimates and quasilinear parabolic equations. Comm. Pure Appl. Math. **21**, 205–226 (1968)
40. K. Uhlenbeck, Regularity for a class of nonlinear elliptic systems. Acta Math. **138**, 219–240 (1977)
41. J.M. Urbano, *The Method of Intrinsic Scaling. A Systematic Approach to Regularity for Degenerate and Singular PDEs*. Lecture Notes in Mathematics, vol. 1930 (Springer, Berlin, 2008)

A Short Presentation of Emmanuele's Work

Daniele Andreucci and Ugo Gianazza

Abstract As the title suggests, this is a short presentation of DiBenedetto's mathematical work. In the first part Ugo Gianazza gives a general overview, without entering too much into details of specific papers; in his contribution Daniele Andreucci focuses on DiBenedetto's accomplishments in BioMathematics.

Keywords Biomathematics · Phototransduction · Stefan problem · DeGiorgi's class · Intrinsic scaling · Semilinear elliptic equation

1 The Beginning

1.1 Master Degree and an Encounter with Carlo Pucci

Emmanuele was born in Lentini, an ancient Greek town in Sicily, founded by colonists from Naxos as Leontini in 729 BC. Even though Lentini is basically halfway between Syracuse and Catania, and this latter city is the home of a famous and important university in Southern Italy, nevertheless, following his parents' suggestions, after finishing high school, Emmanuele chose the University of Florence, where he enrolled in the Mathematics program. He studied, and at the same time he worked as a waiter in local restaurants, so that he could be economically independent.

Emmanuele discussed his *tesi di laurea* (similar, but not exactly comparable to a master thesis) in 1975 at the Department of Mathematics of the University of Florence. The title was "Soluzioni deboli per l'equazione della diffusione nei mezzi

D. Andreucci (✉)
Dipartimento di Scienze di Base e Applicate per l'Ingegneria, Sapienza Università di Roma, Roma, Italy
e-mail: daniele.andreucci@sbai.uniroma1.it

U. Gianazza
Dipartimento di Matematica "F. Casorati", Università di Pavia, Pavia, Italy
e-mail: gianazza@imati.cnr.it

V. Vespri et al. (eds.), *Harnack Inequalities and Nonlinear Operators*, Springer INdAM Series 46, https://doi.org/10.1007/978-3-030-73778-8_2

porosi," (Weak solutions to the diffusion equation in porous media). Advisor and co-advisor were Mario Primicerio, and Carlo Pucci, respectively.

Some time later, in the corridors of the Department, Emmanuele met Pucci, who had in his hands a copy of a CNR (Italy National Research Council) call for US grants for Italian students. With his typical straightforward way of doing, Pucci did not spend too many words and simply said: "DiBenedetto, faccia domanda," (DiBenedetto, apply).

Emmanuele applied, got this grant and left, destination University of Texas at Austin.

1.2 The Ph.D.

In Texas Emmanuele worked very hard (setting the standard, that all his future PhD students would have to comply with), and eventually got his Ph.D. at Austin in 1979. The title of his dissertation was "Implicit Degenerate Evolution Equations in Hilbert Spaces." Ralph E. Showalter was the advisor. The results appeared in a joint paper with Showalter, which was published in 1981 [1].

From the very beginning, Emmanuele was autonomous in his research work; in particular, one of the issues he started thinking about, was the regularity of the solutions of the problems he had become familiar with, working with Primicerio, Showalter, Cannon.

That is why he ended up studying by himself the three different approaches to regularity, namely

- DeGiorgi's Method,
- Moser's Method,
- Nash's Method,

and, in particular with the first one, re-visiting them in his own way.

2 The First Works

2.1 The Stefan Problem ...

Even though there are some previous papers of his about the regularity of solutions to the porous medium equation, a topic Emmanuele has repeatedly come back to in his career, the most important accomplishment of that initial period is the paper [2].

In my opinion the MathSciNet-review is an example of understatement:

The author considers parabolic equations of the form

$$\partial_t \beta(u) - \operatorname{div} \mathbf{a}(x, t, u, \nabla_x u) + b(x, t, u, \nabla_x u) \ni 0,$$

where $\beta(\cdot)$ represents a maximal monotone graph in $\mathbb{R} \times \mathbb{R}$ such that $0 \in \beta(0)$. Under reasonable assumptions on **a**, *b and β, he demonstrates that any weak solution u which is essentially bounded with $\partial u/\partial t \in L^2$ is continuous. The method of proof employs the techniques and results of O. A. Ladyzhenskaya, V. A. Solonnikov and N. N. Ural'tseva.*

Here DiBenedetto deals with the case in which $\beta(\cdot)$ has a jump at the origin. More precisely, he assumes that $\beta(\cdot)$ is given by

$$\beta(r) = \begin{cases} \beta_1(r) & r > 0 \\ [-\nu, 0] & r = 0 \\ \beta_2(r) - \nu & r < 0 \end{cases}$$

where $\nu > 0$ is a given constant and $\beta_i(\cdot)$, $i = 1, 2$, are monotone increasing functions in their respective domain of definition, a.e. differentiable and

$$0 < \alpha_o \leq \beta_i'(r) \leq \alpha_1 \qquad i = 1, 2$$

for two positive constants α_o, α_1, and $\beta_1(0) = \beta_2(0) = 0$.

The continuity of weak solutions of these parabolic PDEs had been a long standing open problem, at which many talented mathematicians had tried their hands, and here comes the solution by a young researcher, under quite general assumptions on β and on the operator. The *techniques and results of O. A. Ladyzhenskaya, V. A. Solonnikov, and N. N. Ural'tseva*, mentioned in the review, are nothing but a novel adaptation to the singular parabolic setting represented by the Stefan Problem, of DeGiorgi's Method.

DiBenedetto considers a maximal monotone graph which exhibits a single finite jump, but it becomes natural to raise the question of a graph β exhibiting multiple jumps and/or singularities of other kind. For these more general graphs, in the mid 1990s, it was established in [3] that solutions are continuous provided $N = 2$. When $N \geq 3$ the same conclusion holds, provided that $a(x, t, u, \nabla_x u)$ reduces to the Laplacian. Several papers have extended and improved these results for specific graphs (see, for example, [4, 5]) or have given precise estimates on the modulus of continuity (see [6, 7]), but, for $N \geq 3$, it remains an open question whether bounded solutions of the above differential inclusion, with a having the full quasilinear structure, and for a general coercive maximal monotone graph β, are locally continuous in their domain of definition.

2.2 ...*and the Porous Medium Equation*

After the previous paper on the Stefan Problem, Emmanuele extended the continuity result to a different kind of diffusion equations in [8]. Namely, he dealt with the case of β continuous, coercive, monotone in $\mathbb{R}$, such that $\beta'(s)$ blows up at $s = 0$. The model example he had in mind was

$$\beta(s) = |s|^{1/m} \operatorname{sign} s, \qquad m > 1,$$

which typically occurs when studying the filtration of gases in porous medium, and the flow obeys a polytropic regime.

The method of proof closely resembled the one of [8], even though a different analysis had to be produced in order to take into account the particular nature of the singularity of β; this affects the modulus of continuity for u. DiBenedetto would later come back to this same problem, and would dramatically improve the modulus of continuity, relying on the so-called *intrinsic scaling* (see below, but also the contribution by Vincenzo Vespri).

With [8], the general perception of Emmanuele's work started changing. Its MathSciNet-review begins with: *In this important paper the author establishes a modulus of continuity after a very broad class of equations modelled after the porous medium equation. A significant consequence of the modulus of continuity result is that a family of equibounded solutions is always relatively compact in the class of continuous functions with uniform norm …*

3 A Brief Overview

This is not the place to systematically study Emmanuele's overall scientific production (also because it is not completed yet, since DiBenedetto is still actively doing research). Nevertheless, let me just briefly focus on two important aspects of his work.

3.1 DeGiorgi's Classes, Both Elliptic and Parabolic

If we restrict ourselves for simplicity to the elliptic context, let E be an open subset of $\mathbb{R}^N$ and for $y \in \mathbb{R}^N$, let $K_\rho(y)$ denote a cube of edge 2ρ centered at y. The DeGiorgi classes $[DG]_p^\pm(E;\gamma)$ in E are the collection of functions $u \in W_{loc}^{1,p}(E)$, for some $p > 1$, satisfying

$$\int_{K_\rho(y)} |D(u-k)_\pm|^p dx \le \frac{\gamma}{(R-\rho)^p} \int_{K_R(y)} |(u-k)_\pm|^p dx \qquad (1.1)_\pm$$

for all cubes $K_\rho(y) \subset K_R(y) \subset E$, and all $k \in \mathbb{R}$, for a given positive constant γ.

Nobody can say what exactly DeGiorgi saw in his mind, when he introduced them, and developed his approach to regularity. He probably envisioned that they had a broader applicability, and he actually came close to saying something of this sort, when he discussed his conjectures about linear elliptic equations with unbounded coefficients (see [9]).

On the other hand, the original DeGiorgi's paper [10] is hard to read, and even the fundamental re-writing by the Russian School (see [11, 12]) leaves the main difficulties clearly in sight.

By repeatedly applying them in different contexts, Emmanuele has clearly shown that DeGiorgi's classes are an extremely powerful tool for regularity, and paved the way to their systematic use in the study of regularity.

As a matter of fact, it is not only a question of technical method: proving with N. Trudinger in [13] that functions in DeGiorgi classes satisfy the Harnack inequality, shows that this property has nothing to do with PDEs (at least in the elliptic setting), but it is just a Real Analysis fact.

Chances are that the same could be true also for the Wiener criterion for continuity at the boundary, but this is another story, which is still at its very beginning (see, for example [14, 15]).

3.2 The Method of Intrinsic Scaling

The technique, which Emmanuele developed with different collaborators (Friedman, Chen, Manfredi, Kwong, Vespri, etc.) has introduced a new way to look at the regularity properties of degenerate and singular parabolic equations. Listing all the accomplishments in this field, is obviously impossible; in his contribution Vincenzo Vespri will discuss some of the features of this issue.

This overview gives me the opportunity to highlight a typical Emmanuele's feature: after developing this tool, he did not keep it as a personal possession, that nobody else should be allowed to employ; it was quite the opposite: he has always been extremely happy that others could use it, and apply it to problems he had never thought about. Moreover, he has always been very generous, ready both to give any sort of advice when asked, and to praise someone's else achievements.

4 A More Specific (but Rather Quick) Look

Now I would like to discuss three papers by DiBenedetto, which are perhaps less known, at least when compared with others works of his, but which nonetheless represent important achievements.

4.1 A Semilinear Elliptic Equation

In [16] DiBenedetto and Diller consider the semilinear equation in $L^1(\mathbb{R}^N)$,

$$\lambda u - \Delta \ln u = f, \qquad \lambda > 0, \tag{1}$$

where $f \in L^1_{\mathrm{loc}}(\mathbb{R}^N)$ is non-negative, and $\lambda > 0$. We say that (u, f) is a solution pair to (1), if $u : \mathbb{R}^N \to \mathbb{R}$ is non-negative, u and $\ln u$ are in $L^1_{\mathrm{loc}}(\mathbb{R}^N)$ and (1) holds in $\mathcal{D}'(\mathbb{R}^N)$. For $N = 2$, they prove that the condition

$$\|f\|_{1,\mathbb{R}^2} > 4\pi$$

is both necessary and sufficient to insure the existence of a nonnegative $L^1_{\mathrm{loc}}(\mathbb{R}^2)$ solution u, for which $\ln u$ is also $L^1_{\mathrm{loc}}(\mathbb{R}^2)$.

Some existence and nonexistence results are also given for the case $N \geq 3$, assuming proper symmetry conditions on f, but we still lack a satisfactory existence condition for a general f.

The problem is interesting in itself, but also because it can be seen as the implicit Euler discretization of the logarithmic diffusion equation

$$\partial_t u - \Delta \ln u = 0 \quad \text{in} \quad \mathbb{R}^N \times (0, T), \quad u(\cdot, 0) = u_o$$

for some given initial datum u_o (see [17]). Such an equation arises in the study of the two-dimensional Ricci flow, and in connection with the dynamics of thin liquid films.

4.2 *The* Δ_∞

In [18] the problem is to investigate the limiting behavior as $p \to \infty$ of the family of p-Laplacian problems

$$\begin{cases} \operatorname{div}(|\nabla u_p|^{p-2}\nabla u_p) = -f, \\ u_p \in W^{1,p}_o(\Omega), \end{cases} \tag{2}$$

in a bounded simply connected domain Ω in $\mathbb{R}^N$ with smooth boundary, where f is a given nonnegative function in Ω.

The limiting solutions u_∞ satisfy

$$|\nabla u_\infty| = 1 \quad \text{in the set } [f > 0] \text{ in the viscosity sense,}$$

$$\Delta_\infty u_\infty = 0 \quad \text{in the set } \overline{[f > 0]}^c \text{ in the viscosity sense,}$$

where

$$\Delta_\infty v := \sum_{i,j=1}^{N} v_{x_i} v_{x_j} v_{x_i x_j}.$$

Physically, when $N = 2$, the solution corresponds to the stress potential function for an ideal plastic state of a rod subject to a torsional moment for an extended time at high temperature.

The sequence of unconstrained variational problems (2) posed in a uniformly convex Banach space yields a constrained extremal problem in a non-reflexive Banach space.

4.3 A Theorem by DiBenedetto, Friedman and … Newton

Let K be a domain in $\mathbb{R}^N$ containing the origin, and let $\lambda > 1$. Then the shell $\lambda K \setminus K$ is called a homoeoid with center at the origin

Newton (see [19, pag. 22]) proved that if K is an ellipsoid with center at the origin, then the gravity force produced by a uniform distribution of mass on the homoeoid $\lambda K \setminus K$ is equal to zero at any point of K.

It is quite natural to ask whether the converse is true as well, namely: Are ellipsoids the only bodies having the property that the gravitational force induced by a homoeoid is zero at all internal points?

A positive answer for $N = 3$ was given by Dive in[20], Emmanuele and Friedman proved it for any $N \geq 2$ in [21, Theorem 5.1].

5 Applications to Biology

The work of Emmanuele in Biomathematics may be seen as an expression of his belief that physical motivation is important in Mathematics, and that vice-versa the applied sciences may greatly benefit from the rigorous approach of Mathematical Physics. From this point of view, let me quote the handbook of Rational Mechanics [22], as a very pertinent example of Emmanuele's look at natural science, that is precise, terse and deep. The book grew out of his notes of the course he gave for many years in Italy at the University of Rome Tor Vergata, in the 1980s and 1990s of last century.

It was in fact with the frame of mind of the mathematical physicist that Emmanuele around the year 2000 started a project about mathematical modeling of phototransduction in the eye of vertebrates. The idea of undertaking this effort, as far as I know, dated to some years before and was of course related to discussions with his wife Heidi Hamm, a distinguished scientist in the field of cell signaling and of phototransduction. The physical problem is quite complex and the mathematical modeling of it accordingly tackled many issues. We will review some, as a section of the results. Phototransduction is the process allowing the retina to transform the signal carried by a photon into an electric signal to be eventually conveyed to the brain. The first stages of the process take place in the photoreceptors cells, rods and cones, and this was the section of interest.

Firstly, the geometrical intricacy of the receptor cells called for some kind of complexity reduction, which was obtained via simultaneous application of homogenization theory and concentration of capacity. The mathematical paper [23] was published on this topic, but the mathematics was developed to solve the real world problem, keeping the principles recalled elsewhere in this Introduction, and the model appeared in applicative journals, like the Biophysical Journal [24]. As further witness to this approach, let me recall that advances in modeling and mathematics were always either lead or immediately followed by the numerical analysis of the model and by the comparison of the results with the experimental data available in the literature. Also, the homogenized model was thoroughly compared to the 'fully resolved' model in [25] both in terms of gain in computation time and of accuracy.

Another unavoidable difficulty in the phenomenon is the random character of the photon event triggering vision, causing variability in the response of the cell, whence the relevant issue of reduction of such variability (Fig. 1). This issue is connected also to the geometry of the cells, for example via the presence of incisures, which are clefts in the disks filling the rod outer segment, see [26, 27]. This is an instance of one of the motivations of the modeling effort, that is the attempt to give proper account of spatial resolution in the diffusion process, avoiding the shortcut of treating the cell as a well-stirred droplet. Indeed it was shown in [28] that the diffusion cascade transporting the signal inside the cell does have a role in the reduction of variability. The relevance for phototransduction of diffusion on

Fig. 1 Light and shadow, or the importance of variability in visual transduction (Nashville, December 2005)

membranes and in the cytosol, and more generally of morphology, is also addressed in [29, 30].

In his thorough view of applications of Mathematics to Biology Emmanuele also lead the estimation of the sensitivity of the model to the variation of physical parameters, an issue whose importance is well understood by everybody working in the field, where experimental data are often difficult to find and interpret, see [31].

5.1 *Other Topics*

The contribution of Emmanuele to Applied Mathematics is by no means limited to visual transduction. Fluorescence recovery after photobleaching (FRAP) is an experimental technique employed to measure the diffusivity of molecules in cells. The derivation of a reliable estimate of the target coefficient from raw data is however far from being trivial and involves a good deal of mathematical modeling, see [32, 33]. A reaction-diffusion model based on experimental data recovered by FRAP was introduced in [34], in order to model the behavior of p53, a protein connected to tumor suppression.

Other work is mostly connected to signaling, see e.g. [35]. Let us conclude however this very short review by quoting the paper [36], where Emmanuele returns to one of his main points of interest, the connection between diffusion and geometry, investigating surface diffusion along tubular structures and quantitatively comparing it to diffusion on flat surfaces. Tubular structures are relevant in the biology of the cell, and the paper shows that their geometry may have an important effect in slowing down diffusion of various species.

6 There is Much More

6.1 *Few Numbers*

Even though one cannot pretend that sheer numbers are able to describe a scientific life, nevertheless they can help in gaining a more thorough prospect.

- Emmanuele was the Editor in Chief of the SIAM Journal on Mathematical Analysis from 1991 to 2000 and definitely contributed to its present reputation as one of the leading journal in the field; he was and still is member of the editorial board of many different scientific journals.
- He has authored or edited 12 between books and lecture notes;
- He has advised 6 Ph.D. students;
- He has had 17 PostDocs and Research Associates, even though the number of people, who have benefitted from his scientific input, is much larger; as I have already mentioned above, DiBenedetto has always been very generous in sharing ideas, suggestions, conjectures;

- He has produced 121 publications, quite frequently contributing with the main ideas.

7 Some Personal Words by Ugo Gianazza

7.1 *What have I Learnt?*

It is not easy for me to exhaustively list all the things that I have learned from Emmanuele since the first time, when I met him at the Department of Mathematics at Northwestern University in 1999, where he had invited me. Sticking to the scientific side, here is a coincise summary of what I have received from him (he has surely taught me more, but I am quite sure that simply I did not realize it):

- Problems need to be motivated;
- Never forget the physical motivation of the problem you are studying at the moment; in other words, if you know where the problem comes from, this will also tell you, what kind of properties you should look for;
- Avoid trivial problems, just for the sake of having one paper more;
- Papers should be clearly focused: when the reader starts looking at it, it should be clear from the very beginning what is all about.
- Papers must be written in a terse way, and there should be no more words than strictly necessary to convey the scientific message one is trying to transmit;
- Give credit: for sure, you are not the only one working in the field, a lot of people have contributed to the present knowledge, and everyone needs to be fully and properly recognized;
- No quick compromise: it is easy to be satisfied with an easy solution or a shortcut, but one should consider, whether it is really worth doing it.

7.2 *Just One Final Picture*

In November 2016, I visited Emmanuele at Vanderbilt University in Fall; we went once for a walk with Naian Liao and Colin Klaus, two of his former PhD students, who were also there at the time, and we seized the opportunity and took the picture you can see here. When I was back home, my 5-year-old nephew asked me, where I had been, and telling him who I had visited, and what I had done, it turned out to be quite natural for me to show him the picture. He looked at it, and then quickly commented: “These are your friends, aren’t they?” This struck me, because it reminded me of a sentence I had once read in an interview with DeGiorgi: “Mathematics is something you do with friends.” I think that this precisely describes my overall experience as a mathematician, and in particular my work with Emmanuele (Fig. 2).

Fig. 2 Emmanuele with Ugo Gianazza and two of his former PhD students, Naian Liao and Colin Klaus

7.3 A Final Remark

Let me conclude my contribution with a personal note. There is a college in the system of the University of Pavia, Collegio Ghislieri, founded in 1567, whose original motto is: "Sapientia cum probitate morum coniuncta humanae mentis perfectio", which means "Wisdom together with moral integrity is the perfection of the human mind." In my opinion, it is hard to find a better description of who Emmanuele is and represents for all those who have been lucky to get in touch with him.

References

1. E. DiBenedetto, R.E. Showalter, Implicit degenerate evolution equations and applications. SIAM J. Math. Anal. **12**(5), 731–751 (1981)
2. E. DiBenedetto, Continuity of weak solutions to certain singular parabolic equations. Ann. Mat. Pura Appl. (4) **130**, 131–176 (1982)
3. E. DiBenedetto, V. Vespri, On the singular equation $\beta(u)_t = \Delta u$. Arch. Ration. Mech. Anal. **132**, 247–309 (1995)

4. U. Gianazza, B. Stroffolini, V. Vespri, Interior and boundary continuity of the weak solution of the singular equation $(\beta(u))_t = Lu$. Nonlinear Anal. **56**, 157–183 (2004)
5. U. Gianazza, V. Vespri, Continuity of weak solutions of a singular parabolic equation. Adv. Differ. Equations **8**, 1341–1376 (2003)
6. P. Baroni, T. Kuusi and J.M. Urbano, A quantitative modulus of continuity for the two-phase Stefan problem. Arch. Ration. Mech. Anal. **214**, 545–573 (2014)
7. P. Baroni, T. Kuusi, C. Lindfors, J.M. Urbano, Existence and boundary regularity for degenerate phase transitions. SIAM J. Math. Anal. **50**(1), 456–490 (2018)
8. E. DiBenedetto, Continuity of weak solutions to a general porous medium equation. Indiana Univ. Math. J. **32**(1), 83–118 (1983)
9. E. DeGiorgi, Congetture sulla continuità delle soluzioni di equazioni lineari ellittiche autoaggiunte a coefficienti illimitati, in *Typewritten document, Lecce (Italy), January 4 1995* (1995)
10. E. DeGiorgi, Sulla differenziabilità e l'analiticità delle estremali degli integrali multipli regolari. Mem. Accad. Sci. Torino Cl. Sci. Fis. Mat. Natur. (3) **3**, 25–43 (1957)
11. O.A. Ladyzenskaya, N.A. Solonnikov, N.N. Ural'tzeva, *Linear and Quasilinear Equations of Parabolic Type*. Translations of Mathematical Monographs, vol. 23 (American Mathematical Society, Providence, 1967)
12. O.A. Ladyzenskaya, N.N. Ural'tzeva, *Linear and Quasilinear Elliptic Equations* (Academic Press, New York, 1968)
13. E. DiBenedetto, N. Trudinger, Harnack Inequalities for Quasi-Minima of Variational Integrals. Ann. Inst. Henri Poincaré, Analyse Non Linéaire **1**(4), 295–308 (1984)
14. J. Björn, Sharp exponents and a Wiener type condition for boundary regularity of quasiminimizers. Adv. Math. **301**, 804–819 (2016)
15. E. DiBenedetto, U. Gianazza, A Wiener-type condition for boundary continuity of Quasi-Minima of variational integrals. Manuscripta Math. **149**(3–4), 339–346 (2016)
16. E. DiBenedetto and D.J. Diller, Singular semilinear elliptic equations in $L^1(\mathbb{R}^N)$, in *Topics in Nonlinear Analysis*. Program of Nonlinear Differential Equations Application, vol. 35 (Birkhäuser, Basel, 1999), pp. 143–181
17. E. DiBenedetto, D.J. Diller, About a singular parabolic equation arising in thin film dynamics and in the Ricci flow for complete $\mathbf{R}^2$, in *Partial Differential Equations and Applications*. Lecture Notes in Pure and Application Mathematics, vol. 177 (Dekker, New York, 1996), pp. 103–119
18. T. Bhattacharya, E. DiBenedetto, J.J. Manfredi, Limits as $p \to \infty$ of $\Delta_p u_p = f$ and related extremal problems, in *Some Topics in Nonlinear PDEs (Turin, 1989)*. Rendiconti del Seminario Matematico della Università Politecal, Torino 1989, Special Issue (1991), 15–68
19. O.D. Kellogg, *Foundations of Potential Theory* (Dover, New York, 1953)
20. P. Dive, Attraction des ellipsoïdes homogènes et réciproques d'un théorème de Newton. Bull. Soc. Math. France **59**, 128–140 (1931)
21. E. DiBenedetto, A. Friedman, Bubble growth in porous media. Indiana Univ. Math. J. **35**(3), 573–606 (1986)
22. E. DiBenedetto *Classical Mechanics* (Cornerstones, Birkhäuser, New York, 2011)
23. D. Andreucci, P. Bisegna, E. DiBenedetto, Homogenization and concentrated capacity for the heat equation with non–linear variational data in reticular almost disconnected structures and applications to visual transduction. Annali di Matematica Pura e Applicata **182**, 375–407 (2003)
24. D. Andreucci, P. Bisegna, G. Caruso, H. Hamm, E. DiBenedetto, Mathematical model of the Spatio-Temporal dynamics of second messengers in visual transduction. Biophys. J. **85**, 1358–1376 (2003)
25. G. Caruso, H. Khanal, V. Alexiades, F. Rieke, H. Hamm, E. DiBenedetto, Mathematical and computational modelling of spatio-temporal signalling in rod phototransduction. IEE Proc.-Syst. Biol. **152**, 119–137 (2005)
26. G. Caruso, P. Bisegna, L. Shen, D. Andreucci, H. Hamm, E. DiBenedetto, Modeling the Role of Incisures in Vertebrate Phototransduction. Biophys. J. **91**, 1192–1212 (2006)

27. D. Andreucci, P. Bisegna, E. DiBenedetto, Homogenization and concentration of capacity in the rod outer segment with incisures. Appl. Anal. **85**, 303–331 (2006)
28. P. Bisegna, G. Caruso, D. Andreucci, L. Shen, V.V. Gurevich, H. Hamm, E. DiBenedetto, Diffusion of the second messengers in the cytoplasm acts as a variability suppressor of the single photon response in vertebrate phototransduction. Biophys. J. **94**, 3363–3383 (2008)
29. X.-H. Wen, L. Shen, R.S. Brush, N. Michaud, M.R. Al-Ubaidi, V.V. Gurevich, H.E. Hamm, J. Lem, E. DiBenedetto, R.E. Anderson, C.L. Makino, Overexpression of Rhodopsin alters the structure and photoresponse of rod photoreceptors. Biophys. J. **96**, 939–950 (2009)
30. C.L. Makino, X.-H. Wen, N.A. Michaud, H.I. Covington, E. DiBenedetto, H.E. Hamm, J. Lem, G. Caruso, Rhodopsin expression level affects rod outer segment morphology and photoresponse kinetics. PLoSONE **7**, e37832 1–7 (2012)
31. L. Shen, G. Caruso, P. Bisegna, D. Andreucci, V.V. Gurevich, H. Hamm, E. DiBenedetto, Dynamics of mouse rod phototransduction and its sensitivity to variation of key parameters. IET Syst. Biol. **4**, 12–32 (2010)
32. M. Kang, C.A. Day, K. Drake, A.K. Kenworthy, E. DiBenedetto, A generalization of theory for two-dimensional fluorescence recovery after photobleaching applicable to confocal laser scanning microscopes. Biophys. J. **97**, 1501–1511 (2009)
33. M. Kang, C.A. Day, A.K. Kenworthy, E. DiBenedetto, Simplified equation to extract diffusion coefficients from confocal FRAP data. Traffic **13**, 1589–1600 (2012)
34. P. Hinow, C.E. Rogers, C.E. Barbieri, J.A. Pietenpol, A.K. Kenworthy, E. DiBenedetto, The DNA binding activity of p53 displays reaction-diffusion kinetics. Biophys. J. **91**, 330–342 (2006)
35. J.N. McLaughlin, L. Shen, M. Holinstat, J.D. Brooks, E. DiBenedetto, H.E. Hamm, Functional selectivity of G protein signaling by agonist peptides and thrombin for the protease-activated receptor-1. J. Biol. Chem. **280**(26), 25048–25059 (2005)
36. C.J.S. Klaus, K. Raghunathan, E. DiBenedetto, A.K. Kenworthy, Analysis of diffusion in curved surfaces and its application to tubular membranes. Mol. Biol. Cell **27**, 3937–3946 (2016)

Evolutionary Problems in Non-Cylindrical Domains

Verena Bögelein, Frank Duzaar, and Christoph Scheven

To Emmanuele DiBenedetto on his 70th birthday

Abstract This survey article presents an existence theory developed in Bögelein et al. (SIAM J Math Anal 50(3): 3007–3057, 2018) for vector-valued gradient flows of integral functionals in bounded non-cylindrical domains $E \subset \mathbb{R}^n \times [0, T)$. The associated system of differential equations takes the form

$$\partial_t u - \operatorname{div} D_\xi f(x, u, Du) = -D_u f(x, u, Du) \qquad \text{in } E,$$

for an integrand $f(x, u, Du)$ that is convex and coercive with respect to the $W^{1,p}$-norm for $p > 1$.

Keywords Parabolic systems · Variational solutions · Noncylindrical domains · Existence · Continuity

1 History of the Problem

The largest part of mathematical literature on evolutionary problems deals with cylindrical domains. This means that the parabolic differential equation is prescribed in a space-time cylinder $\Omega \times (0, T)$ with a spatial domain $\Omega \subset \mathbb{R}^n$ and $T \in (0, \infty]$. However, in many applications like fluid dynamics [21, 29], modeling

V. Bögelein (✉)
Fachbereich Mathematik, Universität Salzburg, Salzburg, Austria
e-mail: verena.boegelein@sbg.ac.at

F. Duzaar
Department Mathematik, Universität Erlangen–Nürnberg, Erlangen, Germany
e-mail: duzaar@math.fau.de

C. Scheven
Fakultät für Mathematik, Universität Duisburg-Essen, Essen, Germany
e-mail: christoph.scheven@uni-due.de

V. Vespri et al. (eds.), *Harnack Inequalities and Nonlinear Operators*,
Springer INdAM Series 46, https://doi.org/10.1007/978-3-030-73778-8_3

of glacier formations [30] and mathematical biology, in particular in reaction-diffusion systems appearing in pattern formation [14, 26], the problem is modeled by parabolic differential equations in non-cylindrical domains. For more details, we refer to the survey article [20]. Considering evolutionary problems in non-cylindrical domains is also interesting from the mathematical point of view, since new phenomena may appear. This subject has attracted a lot of attention in the last years, cf. [9, 12, 27]. In particular for linear equations there is an extensive literature on the existence theory of such problems, cf. [1, 3, 10, 13, 24, 25] and the references therein. Also regularity issues for evolutionary problems in non-cylindrical domains inspired a lot of research, see for instance [2, 4, 11, 16, 18].

Already De Giorgi became interested in evolutional problems in non-cylindrical domains when he conjectured [15] that a modification of the method of *Minimizing Movements* is a proper tool to show existence of weak solutions to the boundary value problem for the heat equation

$$\partial_t u - \Delta u = 0$$

in non-cylindrical domains $E \subset \mathbb{R}^n \times [0, \infty)$. In the classical case when $E = \Omega \times [0, \infty)$ is cylindric, the method can be explained as follows: first, one considers a time discretization by dividing the time interval $[0, \infty)$ into small intervals of length $h > 0$. On the intervals $(t_{i-1}, t_i]$, where $t_i := ih$ for $i \in \{-1, 0, \dots\}$, one iteratively constructs a sequence $(u_{h,i})_{i=1}^{\infty}$ of minimizers to related elliptic variational problems. More precisely, suppose that for some $i \in \mathbb{N}$

$$u_{h,i-1} \in W_0^{1,2}(\Omega)$$

has already been chosen. If $i = 1$, we let $u_{h,0} = u_o$ be the initial value. Then, we select $u_{h,i}$ as the minimizer of the functional

$$\mathbf{F}_{h,i}(v) := \frac{1}{2} \int_\Omega |\nabla v|^2 \mathrm{d}x + \frac{1}{2h} \int_\Omega |u_{h,i-1} - v|^2 \mathrm{d}x$$

in the class of functions $v \in W_0^{1,2}(\Omega)$. Such a minimizer exists due to the Direct Method of the Calculus of Variations. Next, we define the piecewise in time constant function

$$u_h \colon \Omega \times (-h, \infty) \to \mathbb{R}$$

by

$$u_h(t) := u_{h,i} \quad \text{for } t \in (t_{i-1}, t_i] \text{ where } i \in \mathbb{N}_0.$$

Finally, one can show that in the limit $h \downarrow 0$ the sequence (u_h) converges to a solution of the heat equation on the space-time cylinder $\Omega \times [0, \infty)$ with initial values u_o and homogeneous lateral boundary values.

In the case of a non-cylindrical domain $E \subset \mathbb{R}^n \times [0,\infty)$, De Giorgi [15, Problem 2.2] suggested to consider instead of $\mathbf{F}_{h,i}$ the penalized functional

$$\widetilde{\mathbf{F}}_{h,i}(v) := \mathbf{F}_{h,i}(v) + \log\left(\tfrac{1}{h}\right) \int_{t_{i-1}}^{t_i} \int_{\mathbb{R}^n \setminus E^t} |v|^2 \mathrm{d}x\mathrm{d}t, \tag{1.1}$$

on the function space $W^{1,2}(\mathbb{R}^n)$, where $E^t := \{x \in \mathbb{R}^n : (x,t) \in E\}$ denotes the time slice at time $t \geq 0$. He conjectured that the Minimizing Movements Method applied with the penalized functional $\widetilde{\mathbf{F}}_{h,i}$ leads to a solution of the initial boundary value problem

$$\begin{cases} \partial_t u - \Delta u = 0 \text{ in } E, \\ u = 0 & \text{on } \partial E^t \times (0,\infty), \\ u(\cdot,0) = u_o & \text{in } E^0 \end{cases} \tag{1.2}$$

in the non-cylindrical domain E. To be more precise, De Giorgi considered the inhomogeneous heat equation. For expository reasons we omit the right-hand side here. This conjecture of De Giorgi has been positively answered by Gianazza and Savaré in [17] for the case where E is a relatively open subset of $\mathbb{R}^n \times [0,\infty)$ and $\{E^t\}_{t\in[0,\infty)}$ is a non-decreasing family of open sets. More precisely, they used this technique to show existence of weak solutions to linear parabolic equations in an abstract setting. As an application, their result covers the case of non-cylindrical domains that are non-decreasing in time, i.e.

$$E^s \subseteq E^t \quad \text{for any } 0 \leq s < t \tag{1.3}$$

and thereby verifies De Giorgi's conjecture in this case. Gianazza and Savaré obtained weak solutions

$$u \in C^0\big([0,\infty); L^2(\mathbb{R}^n)\big) \cap L^\infty\big(0,\infty; W^{1,2}(\mathbb{R}^n)\big) \quad \text{with} \quad \partial_t u \in L^2\big(\mathbb{R}^n \times (0,\infty)\big)$$

satisfying

$$\operatorname{spt} u(\cdot,t) \subset \overline{E^t}, \quad \forall t > 0.$$

The latter property guarantees that the homogeneous lateral boundary condition $(1.2)_2$ is satisfied. In the setting of non-cylindrical domains, the fact that the solution belongs to the function space $C^0([0,\infty); L^2(\mathbb{R}^n))$ should already be interpreted as a regularity property ensuring a certain kind of continuity across the boundary with respect to the time direction. One of the main difficulties in the proof of De Giorgi's conjecture was to show that the penalization term forces the limit function to satisfy the homogeneous lateral boundary condition $(1.2)_2$. However, this technique seems to work only for domains that are non-decreasing in time.

In fact, it is plausible that the case of shrinking domains is significantly more difficult because the boundary value problem might become overdetermined if the domain shrinks too fast. In the subsequent paper [28], Savaré succeeded in treating domains whose time slices are uniformly $C^{1,1}$ open sets that possibly decrease at a bounded speed. More precisely, in [28, Theorem 4.4] he obtained an existence result for linear parabolic equations in non-cylindrical relatively open domains $E \subset \mathbb{R}^n \times [0, T]$ for $T > 0$, provided that E_t is an open set of class $C^{1,1}$ and the excess

$$\mathbf{e}(E^s, E^t) := \sup_{x \in E^s} \mathrm{dist}(x, E^t)$$

satisfies

$$\mathbf{e}(E^s, E^t) \leq \int_s^t \rho(\sigma)\mathrm{d}\sigma, \qquad \text{for } 0 \leq s \leq t \leq T, \tag{1.4}$$

for some function $\rho \in L^2(0, T)$. Note that (1.4) only prevents the domain from shrinking too fast. It may still grow arbitrarily fast as in the result by Gianazza and Savaré in [17], since assumption (1.3) implies $\mathbf{e}(E^s, E^t) = 0$ for any $0 \leq s \leq t \leq T$. However, continuity in time across the boundary, i.e. $u \in C^0([0, T]; W^{1,2}(\mathbb{R}^n))$ is only obtained under the additional assumption that condition (1.4) is satisfied for the Hausdorff distance instead of the excess. This means that the domain is neither allowed to shrink too fast, nor to grow too fast. Although this extra assumption seems to be artificial, a similar phenomenon occurs for a uniqueness result proved by a completely different technique, see Sect. 4.3. In [9], Bonaccorsi and Guatteri obtained related results under a slightly different assumption on the domain. They replaced assumption (1.4) by a Hölder condition on the excess $\mathbf{e}(E^s, E^t)$.

The previously mentioned results are limited to the case of linear equations. Nonlinear operators have first been considered by Paronetto [27]. He introduces a different kind of condition on the time-dependent domain by assuming that the time slices are bounded Lipschitz domains that are regular deformations of each other. In a subsequent work, Calvo, Novaga, and Orlandi [12] proved existence of weak solutions to parabolic p-Laplace type equations on bounded domains that are Lipschitz regular in space and time. For the proof of uniqueness, they additionally rely on the assumptions in [27]. The idea of Calvo, Novaga, and Orlandi was to perform a time slicing of height $h_k = \frac{T}{k}$, with $k \in \mathbb{N}$, of the non-cylindrical domain and to solve parabolic equations on the cylindrical domains $E^{ih_k} \times [ih_k, (i+1)h_k)$ of height h_k. These solutions are glued together to a function u_{h_k} defined on $\cup_{i=0}^{k-1} E^{ih_k} \times [ih_k, (i+1)h_k)$. In the limit $k \to \infty$ the sequence (u_{h_k}) converges to a solution of the parabolic equation on the non-cylindrical domain. The main novelty in this paper was the idea to solve parabolic equations on the sliced domains instead of elliptic ones. For possibly decreasing domains this procedure seems to be much more flexible. In fact, it was our starting point towards an existence theory for non-cylindrical domains with almost no restrictions on the domain, see Sect. 4.1.

Before we explain this result in more detail, we will first introduce the general setting and discuss the simpler case of non-decreasing non-cylindrical domains.

2 Gradient Flows in Non-Cylindrical Domains

In the following we will consider boundary value problems for a very general class of vector-valued gradient flows in non-cylindrical domains E. We will present an existence result under the only assumption that E is a bounded and relatively open subset of $\mathbb{R}^n \times [0, T)$ for $n \geq 2$ and some $T > 0$ satisfying

$$\mathcal{L}^{n+1}(\partial E) = 0 \quad \text{and} \quad E \subset \Omega_T := \Omega \times [0, T) \tag{2.1}$$

for some bounded and open set $\Omega \subset \mathbb{R}^n$. For initial data

$$u_o \in L^2\big(E^0, \mathbb{R}^N\big)$$

with $N \geq 1$ and an integrand $f: E \times \mathbb{R}^N \times \mathbb{R}^{Nn} \to [0, \infty]$ satisfying (2.3) below we consider the gradient flow for the associated integral functional, which can formally be written as

$$\begin{cases} \partial_t u - \operatorname{div} D_\xi f(x, u, Du) = -D_u f(x, u, Du) & \text{in } E, \\ u = 0 & \text{on } \displaystyle\bigcup_{t \in [0,T)} \partial E^t \times \{t\}, \\ u(\cdot, 0) = u_o & \text{in } E^0. \end{cases} \tag{2.2}$$

As in the previously mentioned results, we restrict ourselves to the model case of vanishing lateral boundary values. However, we expect that the methods presented below can also be applied to nonzero lateral boundary values, and even time-dependent lateral boundary values.

As integrand we allow any Carathéodory function $f: \Omega \times \mathbb{R}^N \times \mathbb{R}^{Nn} \to [0, \infty]$ fulfilling the convexity and coercivity assumptions

$$\begin{cases} \big(\mathbb{R}^N, \mathbb{R}^{Nn}\big) \ni (u, \xi) \mapsto f(x, u, \xi) \text{ is convex for a.e. } x \in \Omega, \\ \nu |\xi|^p \leq f(x, u, \xi) \text{ for a.e. } x \in \Omega \text{ and all } (u, \xi) \in \mathbb{R}^N \times \mathbb{R}^{Nn}, \end{cases} \tag{2.3}$$

for some constant $\nu > 0$ and some integrability exponent $1 < p < \infty$. Furthermore, we assume that there exists a measurable function $g: \Omega \times [0, \infty) \to [0, \infty)$ such that for any fixed $M \geq 0$ the partial map $g(\cdot, M)$ is integrable, i.e. $g(\cdot, M) \in L^1(\Omega)$, and the pointwise bound

$$0 \leq f(x, u, \xi) \leq g(x, M) \quad \text{for a.e. } x \in \Omega, \text{ provided } |u| \leq M, |\xi| \leq M, \tag{2.4}$$

holds true. Note that this assumption on the integrand f ensures that smooth functions have finite energy, i.e. $f(x, u, Du)$ is integrable whenever u is smooth. The assumption, however, is not very restrictive, since it covers any reasonable growth condition of the form

$$0 \le f(x,u,\xi) \le \alpha(x)\phi(|\xi|) + \beta(x)\psi(|u|) + \gamma(x)$$

with non-negative continuous functions $\phi, \psi\colon [0,\infty) \to [0,\infty)$ and integrable functions $\alpha, \beta, \gamma \in L^1(\Omega)$. We emphasize that we do not require a p-growth condition from above for the mapping $\xi \mapsto f(x,u,\xi)$, i.e. we assume only p-coercivity as in (2.3)$_2$ and condition (2.4).

In the following, we will work in the framework of so-called variational solutions. Before we state their definition, we shall introduce some notation. As before, for a fixed time $t \in [0,T)$ we denote the section of E by

$$E^t := \{x \in \mathbb{R}^n : (x,t) \in E\} \subset \mathbb{R}^n,$$

so that

$$E \equiv \bigcup_{t \in [0,T)} E^t \times \{t\}.$$

In the following, we denote by V_t the space of those maps $v \in W_0^{1,p}(\Omega,\mathbb{R}^N)$ that vanish outside of E^t, i.e.

$$V_t := \Big\{ v \in W_0^{1,p}(\Omega,\mathbb{R}^N) : v = 0 \text{ a.e. on } \Omega \setminus E^t \Big\}.$$

Note that $V_t \cong W_0^{1,p}(E^t,\mathbb{R}^N)$ if E^t satisfies a measure density condition; see [8, Section 3.2]. Moreover, we will use the abbreviations

$$V^p(E) := \Big\{ u \in L^p\big(0,T; W_0^{1,p}(\Omega,\mathbb{R}^N)\big) : u(t) \in V_t \text{ for a.e. } t \in [0,T) \Big\}$$

and

$$V_2^p(E) := V^p(E) \cap L^2\big(\Omega_T,\mathbb{R}^N\big).$$

By $u(t)$ we denote the function $u(\cdot,t)$. Obviously, the distinction between the two spaces is relevant only in the sub-quadratic case $p < 2$. However, for the sake of a unified treatment of the cases $p \ge 2$ and $p < 2$, we will use the notation $V_2^p(E)$ also in the case $p \ge 2$.

At this point it must be explicitly emphasized that the Dirichlet boundary condition (2.2)$_2$ for the solution u will be encoded in terms of the space $V_2^p(E)$, by simply requiring that $u \in V_2^p(E)$. Note that this ensures only that the Dirichlet boundary values are taken in the horizontal direction. With this respect, ensuring that the solution belongs to the space $C^0([0,T); L^2(\Omega,\mathbb{R}^N))$ should be interpreted as a regularity property.

We now define variational solutions to the initial boundary value problem (2.2) in the non-cylindrical domain E as follows:

Definition 2.1 Suppose that $f\colon \Omega \times \mathbb{R}^N \times \mathbb{R}^{Nn} \to [0, \infty]$ is a Carathéodory function, that $E \subset \Omega \times [0, T)$ is a relatively open non-cylindrical domain and that $u_o \in L^2(E^0, \mathbb{R}^N)$. A measurable map $u\colon \Omega_T \to \mathbb{R}$ in the class

$$u \in L^\infty\big(0, T; L^2(\Omega, \mathbb{R}^N)\big) \cap V_2^p(E)$$

is called a *variational solution* with initial datum u_o if and only if the variational inequality

$$\begin{aligned}\iint_{E\cap\Omega_\tau} f(x, u, Du)\,\mathrm{d}x\,\mathrm{d}t \le \iint_{E\cap\Omega_\tau} \Big[\partial_t v \cdot (v-u) + f(x, v, Dv)\Big]\mathrm{d}x\,\mathrm{d}t \\ - \tfrac{1}{2}\big\|(v-u)(\tau)\big\|^2_{L^2(E^\tau,\mathbb{R}^N)} + \tfrac{1}{2}\big\|v(0) - u_o\big\|^2_{L^2(E^0,\mathbb{R}^N)}\end{aligned} \tag{2.5}$$

holds true, for a.e. $\tau \in [0, T)$ and any $v \in V_2^p(E)$ with $\partial_t v \in L^2\big(\Omega_T, \mathbb{R}^N\big)$.

Variational solutions turn out to be very flexible when proving existence results, since they allow to apply techniques from the Calculus of Variations. Moreover, Definition 2.1 already makes sense for integrands f for which the weak formulation of the associated differential equation does not exist. The notion of variational solution goes back to Lichnewsky and Temam [22] in the context of the time dependent minimal surface equation. The variational inequality (2.5) implies that the variational solution u attains the initial datum u_o in the L^2-sense, see [8, Lemma 3.3]. The proof of this property relies on assumption (2.4), which ensures that a certain mollification of the initial values is a comparison function of finite energy. For a more detailed introduction to variational solutions we refer to [6, 7]. In the following, we will discuss the results obtained in [8] on existence and regularity of variational solutions to gradient flows.

3 Existence and Regularity in Non-Decreasing Domains

We first discuss the easier case of non-decreasing domains, in the sense that the sets E^t do not decrease in time. More specifically, we assume

$$E^s \subseteq E^t \quad \text{for any } 0 \le s \le t \le T. \tag{3.1}$$

Within this context we proved in [8, Theorem 2.3] the existence of a unique variational solution. The precise statement is as follows.

Theorem 3.1 *Suppose that* $f : \Omega \times \mathbb{R}^N \times \mathbb{R}^{Nn} \to [0, \infty]$ *is a variational integrand satisfying* (2.3) *and* (2.4) *and that the non-cylindrical domain* E *satisfies* (2.1) *and* (3.1). *Then, for any initial datum* $u_o \in L^2(E^0, \mathbb{R}^N)$ *there exists a unique variational solution* u *in the sense of Definition* 2.1. *Moreover, the variational solution satisfies*

$$u \in C^0\big([0,T); L^2(\Omega, \mathbb{R}^N)\big) \cap V_2^p(E).$$

We stress that the variational solution constructed in Theorem 3.1 satisfies the regularity property $u \in C^0([0,T); L^2(\Omega, \mathbb{R}^N))$. If the initial datum u_o has finite energy, i.e. if

$$u_o \in V_0 \quad \text{and} \quad \int_\Omega f(x, u_o, Du_o)\, dx < \infty,$$

then we have $\partial_t u \in L^2(\Omega_T, \mathbb{R}^N)$; see [8, Lemma 4.2].

We emphasize that the assumptions in Theorem 3.1 are by far more general than the ones in any of the previously mentioned results. In fact, we are able to deal with minimal assumptions on the domain and the underlying functional, i.e. we essentially only assume $\mathcal{L}^{n+1}(\partial E) = 0$ and convexity and coercivity of the integrand f. Moreover, our methods are flexible enough to include the case of systems.

As proposed by De Giorgi, the construction uses a modification of the method of minimizing movements. However, we omit the penalization term in the elliptic functional (1.1). Instead, we cover the domain by a union of cylinders of height $0 < h \ll 1$ and vary the class of admissible maps from step to step, which enables us to construct minimizers that admit the prescribed boundary values on the boundary of the approximating domain. More precisely, as in the classical case we consider for an integer $\ell \in \mathbb{N}$ a step-size $h_\ell := \frac{T}{\ell}$ and the times $t_{\ell,i} := i h_\ell$ for $i \in \{-1, 0, \dots, \ell\}$. Then, we iteratively construct a sequence of minimizers $(u_{\ell,i})_{i=1}^{\ell}$ to certain elliptic variational integrals. We extend u_o by zero to Ω and let $u_{\ell,0} = u_o$. Suppose that for some $i \in \{1, \dots, \ell\}$

$$u_{\ell,i-1} \in L^2\big(\Omega, \mathbb{R}^N\big)$$

has already been chosen. Then, we select $u_{\ell,i}$ as the minimizer of the elliptic variational functional

$$\mathbf{F}_{\ell,i}(v) := \frac{1}{2}\int_\Omega |\nabla v|^2 dx + \frac{1}{2h_\ell}\int_\Omega |u_{\ell,i-1} - v|^2 dx$$

in the function class

$$v \in L^2\big(\Omega, \mathbb{R}^N\big) \cap V_{t_{\ell,i}}.$$

As in the classical case, we define the piecewise in time constant function $u_\ell\colon \Omega \times (-h_\ell, T] \to \mathbb{R}$ by

$$u_\ell(t) := u_{\ell,i} \quad \text{for } t \in (t_{\ell,i-1}, t_{\ell,i}], \text{ where } i \in \{0, \dots, \ell\}.$$

Exploiting the minimizing property of the functions $u_{\ell,i}$ and taking into account that $v \equiv 0$ is an admissible comparison map, one can show the following energy estimate

$$\sup_{t\in[0,T)} \int_\Omega |u_\ell(t)|^2 \mathrm{d}x + \nu \iint_{\Omega_T} |Du_\ell|^p \mathrm{d}x\mathrm{d}t \le 3T \int_\Omega g(x,0)\,\mathrm{d}x + 2\int_\Omega |u_o|^2 \mathrm{d}x.$$

This ensures that the sequence $(u_\ell)_{\ell\in\mathbb{N}}$ is uniformly bounded in the function spaces $L^\infty(0,T; L^2(\Omega, \mathbb{R}^N))$ and $L^p(0,T; W^{1,p}(\Omega, \mathbb{R}^N))$. Therefore, there exists a limit map

$$u \in L^\infty\big(0,T; L^2(\Omega,\mathbb{R}^N)\big) \cap L^p\big(0,T; W_0^{1,p}(\Omega,\mathbb{R}^N)\big)$$

and a not re-labelled subsequence such that

$$\begin{cases} u_\ell \overset{*}{\rightharpoonup} u \text{ weakly* in } L^\infty\big(0,T; L^2(\Omega,\mathbb{R}^N)\big), \\ u_\ell \rightharpoonup u \text{ weakly in } L^p\big(0,T; W^{1,p}(\Omega,\mathbb{R}^N)\big) \end{cases}$$

in the limit $\ell \to \infty$. By construction we know that $u_\ell \equiv 0$ a.e. in $\Omega_T \setminus E^{(\ell)}$, where

$$E^{(\ell)} := \bigcup_{i=1}^{\ell} Q_{\ell,i}, \quad \text{with } Q_{\ell,i} := E^{t_{\ell,i}} \times [t_{\ell,i-1}, t_{\ell,i}).$$

Since the sets $E^{(\ell)}$ approximate the non-cylindrical domain from the outside, i.e. $E \subset E^{(\ell)}$ for any $\ell \in \mathbb{N}$, we may conclude that the limit function satisfies $u \equiv 0$ a.e. in $\Omega_T \setminus \overline{E}$. By a Fubini argument and due to assumption $(2.1)_1$, this ensures that $u(t) \in V_t$ for a.e. $t \in [0,T)$ and hence $u \in V_2^p(E)$. In the next step, one has to ensure that the limit map u solves the variational inequality (2.5). To this aim, we first observe that for any $\ell \in \mathbb{N}$ and any $\tau \in [0,T)$ the function u_ℓ is a minimizer of the functional

$$\mathbf{F}_\ell(v) := \int_{\Omega_\tau} \Big[f(x,v,Dv) + \tfrac{1}{2h_\ell}|v(t) - u_\ell(t-h_\ell)|^2\Big]\mathrm{d}x\mathrm{d}t$$

in the class of functions $v \in L^\infty(0,T; L^2(\Omega,\mathbb{R}^N)) \cap L^p(0,T; W_0^{1,p}(\Omega,\mathbb{R}^N))$ satisfying $v(t) \in V_{h_\ell \lceil \frac{t}{h_\ell} \rceil}$ for a.e. $t \in [0,T)$. Now, we consider a comparison map $v \in V_2^p(E)$ with $\partial_t v \in L^2\big(\Omega_T, \mathbb{R}^N\big)$. Since $V_s \subset V_t$ holds for $0 \le s \le t$ by hypothesis (3.1), we have that $v(t) \in V_{h_\ell \lceil \frac{t}{h_\ell} \rceil}$ for a.e. $t \in [0, T)$. Therefore, for any $s \in (0, 1)$ the function $v_s = u_\ell + s(v - u_\ell)$ is an admissible comparison map in the functional $\mathbf{F}_\ell$. With the help of the convexity of $(u, \xi) \mapsto f(x, u, \xi)$, we can show that after passing to the limit $s \downarrow 0$, this leads us to the inequality

$$\iint_{\Omega_\tau} f\big(x, u_\ell, Du_\ell\big) \mathrm{d}x\,\mathrm{d}t \le \iint_{\Omega_\tau} f(x, v, Dv) \mathrm{d}x\,\mathrm{d}t + \iint_{\Omega_\tau} \Delta_t^{h_\ell} u_\ell \cdot \big(v - u_\ell\big) \mathrm{d}x\,\mathrm{d}t,$$

where we have abbreviated the finite difference quotient in time by

$$\Delta_t^{h_\ell} u_\ell(t) := \tfrac{1}{h_\ell}\big(u_\ell(t) - u_\ell(t - h_\ell)\big).$$

By lower semi-continuity, we obtain for the left-hand side that

$$\iint_{\Omega_\tau} f(x, u, Du) \mathrm{d}x\mathrm{d}t \le \liminf_{\ell \to \infty} \iint_{\Omega_\tau} f\big(x, u_\ell, Du_\ell\big) \mathrm{d}x\mathrm{d}t.$$

Formally, the second integral on the right-hand side converges to

$$\begin{aligned}\iint_{\Omega_\tau} \partial_t u \cdot (v - u) \mathrm{d}x\,\mathrm{d}t = &\iint_{\Omega_\tau} \partial_t v \cdot (v - u) \mathrm{d}x\,\mathrm{d}t \\ &- \tfrac{1}{2}\|(v - u)(\tau)\|^2_{L^2(\Omega, \mathbb{R}^N)} + \tfrac{1}{2}\|v(0) - u_o\|^2_{L^2(\Omega, \mathbb{R}^N)}.\end{aligned}$$

Since the weak time derivative $\partial_t u$ might not exist, this computation is purely formal. Nevertheless, this computation can be made rigorous by a careful analysis of the involved difference quotients and a Fubini argument. In this way, we end up with the variational inequality

$$\begin{aligned}&\iint_{\Omega_\tau} f(x, u, Du) \mathrm{d}x\,\mathrm{d}t + \tfrac{1}{2}\|(v - u)(\tau)\|^2_{L^2(\Omega, \mathbb{R}^N)} \\ &\quad\le \iint_{\Omega_\tau} \Big[f(\cdot, v, Dv) + \partial_t v \cdot (v - u)\Big] \mathrm{d}x\,\mathrm{d}t + \tfrac{1}{2}\|v(0) - u_o\|^2_{L^2(\Omega, \mathbb{R}^N)}\end{aligned}$$

for a.e. $\tau \in [0, T)$ and any comparison map $v \in V_2^p(E_T)$ with $\partial_t v \in L^2(\Omega_T, \mathbb{R}^N)$. This proves that u is a variational solution in the sense of Definition 2.1. The regularity property that $u \in C^0([0, T); L^2(\Omega, \mathbb{R}^N))$ is shown in [8, § 4.1.6], while the proof of the uniqueness can be found in [8, Lemma 4.1].

4 Variational Solutions in General Variable Domains

4.1 Existence of Variational Solutions

The case of domains possibly shrinking in time is significantly more difficult. In this situation the boundary value problem might be over-determined. Surprisingly, in [8, Theorem 2.6] we were able to prove existence of variational solutions in possibly decreasing domains under otherwise the same conditions as before.

Theorem 4.1 *Suppose that* $f: \Omega \times \mathbb{R}^N \times \mathbb{R}^{Nn} \to [0, \infty]$ *is a variational integrand satisfying* (2.3) *and* (2.4) *and that the non-cylindrical domain E satisfies* (2.1). *Then, for any initial datum* $u_o \in L^2(E^0, \mathbb{R}^N)$, *there exists a variational solution u in the sense of Definition 2.1.*

The reason for the generality of this result is that in Definition 2.1, we impose only little regularity on the variational solution, i.e. we consider solutions in $L^\infty(0, T; L^2(\Omega, \mathbb{R}^N))$, rather than $C^0([0, T); L^2(\Omega, \mathbb{R}^N))$. The advantage of this viewpoint is that existence of variational solutions can be shown for a huge class of non-cylindrical domains. The property that solutions are continuous in time across the boundary, i.e. that they belong to $C^0([0, T); L^2(\Omega, \mathbb{R}^N))$ should in this setting be interpreted as a regularity property. It is natural that for this property and for the uniqueness of solutions further assumptions are required that prevent the domain from shrinking too fast.

In the case of possibly decreasing domains, it turns out that a direct application of the minimizing movements method does not yield an existence result in the generality of Theorem 4.1. Instead, we employ a different strategy. Again, we approximate the non-cylindrical domain from the outside by a union of cylinders of height $0 < h_\ell := \frac{T}{\ell}$, with $\ell \in \mathbb{N}$, defined by

$$Q_{\ell,i} := E_{\ell,i} \times I_{\ell,i},$$

where

$$E_{\ell,i} := \bigcup_{t \in I_{\ell,i}} E^t \subset \mathbb{R}^n \quad \text{and} \quad I_{\ell,i} := [t_{\ell,i-1}, t_{\ell,i}) \text{ with } t_{\ell,i} := i h_\ell$$

for $i \in \{0, \ldots, \ell\}$. On each of the cylinders, we do not solve an elliptic minimization problem as before but, somewhat similar to [12], we use the variational solution of a parabolic system as an approximation. Thereby, we use in each step the final state from the preceding step as new initial values. More precisely, by an inductive procedure we define a sequence $(u_{\ell,i})_{i=1,\ldots,\ell}$ of variational solutions $u_{\ell,i}: Q_{\ell,i} \to \mathbb{R}^N$ defined on the cylindrical domains $Q_{\ell,i}$. For $i = 1$ we let

$$u_{\ell,1} \in C^0\big(\overline{I_{\ell,1}}; L^2(E_{\ell,1}, \mathbb{R}^N)\big) \cap L^p\big(I_{\ell,1}; W_0^{1,p}(E_{\ell,1}, \mathbb{R}^N)\big)$$

be the unique variational solution in the sense of Definition 2.1 from Theorem 3.1 on the cylindrical domain $Q_{\ell,1}$ with initial datum u_o. Note that $E^0 \subset E_{\ell,1}$ by construction and that u_o vanishes on $E_{\ell,1} \setminus E^0$. Supposed that the mapping $u_{\ell,i-1}$ is already defined for some index $i \in \{2, \dots, \ell\}$, then we denote by

$$u_{\ell,i} \in C^0\big(\overline{I_{\ell,i}}; L^2(E_{\ell,i}, \mathbb{R}^N)\big) \cap L^p\big(I_{\ell,i}; W_0^{1,p}(E_{\ell,i}, \mathbb{R}^N)\big)$$

the unique variational solution from Theorem 3.1 on the cylindrical domain $Q_{\ell,i}$ with initial datum

$$w_{\ell,i} := u_{\ell,i-1}(t_{\ell,i-1})\chi_{E_{\ell,i}} \in L^2\big(E_{\ell,i}, \mathbb{R}^N\big)$$

and extend it to $(\Omega \setminus E_{\ell,i}) \times \overline{I_{\ell,i}}$ by zero. At this stage it is crucial that we are able to allow initial values in L^2 as done in Theorem 3.1. Finally, we glue the functions $u_{\ell,i}$ together to form one mapping $u_\ell \colon \Omega_T \to \mathbb{R}^N$ defined on the whole cylinder Ω_T by letting

$$u_\ell(x,t) := u_{\ell,i}(x,t) \quad \text{for } (x,t) \in \Omega \times I_{\ell,i} \text{ with } i \in \{1, \dots \ell\}.$$

Using the fact that each of the functions $u_{\ell,i}$ are variational solutions on the associated cylinders $Q_{\ell,i}$ together with the coercivity of the integrand f, we are able to show the following energy estimate

$$\sup_{t\in[0,T)} \|u_\ell(t)\|^2_{L^2(\Omega,\mathbb{R}^N)} + \nu \iint_{\Omega_T} |Du_\ell|^p \le 5T \int_\Omega g(x,0)\,dx + 2\|u_o\|^2_{L^2(\Omega,\mathbb{R}^N)}.$$

This ensures that the sequence $(u_\ell)_{\ell\in\mathbb{N}}$ is uniformly bounded in the function spaces $L^\infty(0,T; L^2(\Omega,\mathbb{R}^N))$ and $L^p(0,T; W^{1,p}(\Omega,\mathbb{R}^N))$. Therefore, there exists a not re-labelled subsequence and a map

$$u \in L^\infty\big(0,T; L^2(\Omega,\mathbb{R}^N)\big) \cap L^p\big(0,T; W^{1,p}(\Omega,\mathbb{R}^N)\big)$$

such that in the limit $\ell \to \infty$ we have

$$\begin{cases} u_\ell \overset{*}{\rightharpoonup} u \text{ weakly* in } L^\infty\big(0,T; L^2(\Omega,\mathbb{R}^N)\big), \\ u_\ell \rightharpoonup u \text{ weakly in } L^p\big(0,T; W^{1,p}(\Omega,\mathbb{R}^N)\big). \end{cases}$$

Since the union of cylinders $\bigcup_{i=1}^\ell Q_{\ell,i}$ approximates the non-cylindrical domain E from outside, i.e. $E \subset \bigcup_{i=1}^\ell Q_{\ell,i}$ for any $\ell \in \mathbb{N}$, we are able to show that the limit function satisfies $u \equiv 0$ a.e. in $\Omega_T \setminus \overline{E}$. As before, a Fubini argument combined with assumption $(2.1)_1$ ensures that $u(t) \in V_t$ for a.e. $t \in [0,T)$ and hence $u \in V_2^p(E)$. Finally, it remains to pass to the limit in the variational inequality. Again, the notion of variational solutions allows to prove the passage to the limit in the variational inequality only using weak convergence. For the details we refer the reader to [8, § 5.1.4]. This briefly describes the proof of Theorem 4.1.

4.2 Regularity of Variational Solutions

If the domain decreases too fast with respect to time, it is not likely that variational solutions belong to the space $C^0([0,T); L^2(\Omega, \mathbb{R}^N))$. We would like to mention that a related phenomenon has also been observed in [4, 5, 19, 23] by using Petrovskiĭ type constructions. In particular, these constructions yield examples of domains that shrink so fast that solutions of the parabolic p-Laplace equation can become discontinuous in one point of the boundary. Moreover, in case of a jump, i.e. when the domain suddenly becomes smaller, $C^0([0,T); L^2(\Omega, \mathbb{R}^N))$-regularity cannot be expected. These observations suggest that the property $u \in C^0([0,T]; L^2(\Omega, \mathbb{R}^N))$ cannot be obtained for arbitrary domains. Contrary to all previous works, we do not need to restrict ourselves to uniformly $C^{1,1}$ or Lipschitz domains, but we are able to deal with irregular domains whose time slices satisfy a uniform measure density condition. More precisely, we assume that the complements $\mathbb{R}^n \setminus E^t$ of the time slices satisfy the following *measure density condition*: there exists a constant $\delta > 0$ such that for any $t \in [0,T)$ we have

$$\mathcal{L}^n\big((\mathbb{R}^n \setminus E^t) \cap B_r(x_o)\big) \geq \delta \mathcal{L}^n(B_r(x_o)) \quad \text{for any } x_o \in \partial E^t \text{ and } r > 0. \tag{4.1}$$

Note that the measure density condition (4.1) implies that $V_t \cong W_0^{1,p}(E^t, \mathbb{R}^N)$ for any $t \in [0,T)$ (see [8, Lemma 3.1]). Moreover, we of course need an additional assumption that prevents the domain from decreasing too fast. In order to quantify the shrinking of the time slices E^t, Savaré [28] employed the one-sided excess (1.4) and relied on the assumption that the domains E^t are uniformly $C^{1,1}$. In [8] we introduced a new way to measure the decreasing of the domain in time. In fact, our results show that it is more natural to bound the *complementary excess*

$$\mathbf{e}^c(E^s, E^t) := \sup_{x \in \Omega \setminus E^t} \operatorname{dist}(x, \Omega \setminus E^s).$$

We assume the *one-sided growth condition* that there exists a constant $M > 0$ such that there holds

$$\mathbf{e}^c(E^s, E^t) \leq M(t-s) \quad \text{provided } 0 \leq s \leq t < T. \tag{4.2}$$

Note that under the monotonicity assumption (3.1), we have $\mathbf{e}^c(E^s, E^t) = 0$ for any $0 \leq s \leq t < T$. Moreover, we observe that the Lipschitz condition (4.2) on the complementary excess rules out the possibility of a hole developing in the interior of the domain. This phenomenon is excluded in [28] by assuming that the time slices E^t are uniformly $C^{1,1}$. Hence, using the complementary excess allows us to deal with much more general domains that merely satisfy the measure density condition (4.1).

Additionally, we assume that the integrand f satisfies a standard coercivity and growth condition of the form

$$\nu|\xi|^p \le f(x,u,\xi) \le L\big(|\xi|^p + |u|^p + G(x)\big) \tag{4.3}$$

for a.e. $x \in \Omega$ and all $(u,\xi) \in \mathbb{R}^N \times \mathbb{R}^{Nn}$, with some universal structural constants $0 < \nu \le L$ and some $G \in L^{p'}(\Omega_T)$. Note that assumption (4.3) implies $(2.3)_2$ and (2.4). Under these extra assumptions we are able to prove in [8, Theorems 2.7 and 2.8] that the variational solution lies in the space $C^0([0,T]; L^2(\Omega,\mathbb{R}^N))$. For technical reasons we also need the lower bound $p > \frac{2(n+1)}{n+2}$.

Theorem 4.2 *Suppose that $f\colon \Omega\times\mathbb{R}^N\times\mathbb{R}^{Nn} \to [0,\infty]$ is a variational integrand satisfying $(2.3)_1$ and (4.3) for some exponent $p > \frac{2(n+1)}{n+2}$, and that the non-cylindrical domain E satisfies (2.1), (4.1), and (4.2) and that $u_o \in L^2(E^0,\mathbb{R}^N)$. Then, the variational solution u associated to the initial datum u_o obtained in Theorem 4.1 satisfies*

$$u \in C^0\big([0,T]; L^2(\Omega,\mathbb{R}^N)\big),$$

and u attains the prescribed initial values, i.e. $u(0) = u_o$.

The proof of Theorem 4.2 is quite involved. We give only a very brief sketch of the proof and refer to [8] for the details. As a first step, we establish that the variational solution admits a time derivative in the dual space $(V^p(E))'$. This property is proved first for the approximating solutions u_ℓ. For this we have to assume a standard p-growth condition for the functional f. Since the required estimates are uniform with respect to ℓ, the time derivative in $(V^p(E))'$ is preserved in the limit $\ell \to \infty$. Contrary to the cylindrical case, the property $\partial_t u \in (V^p(E))'$ does not immediately yield continuity in time by an embedding theorem. Since we do not know whether such an embedding exists for irregular domains as in Theorem 4.2, we are forced to employ a much more involved strategy.

The key to the proof of continuity in time is an integration by parts formula of the form

$$\langle \partial_t u, \zeta u\rangle \le -\tfrac{1}{2}\iint_E |u|^2 \partial_t \zeta \,\mathrm{d}x\,\mathrm{d}t, \tag{4.4}$$

for any non-negative cut-off function in time $\zeta \in C_0^{0,1}(0,T)$. The fact that we do not obtain an identity in the preceding formula is caused by the one-sided excess assumption (4.2). In fact, under the two-sided assumption (4.6) we get an identity. The proof of the integration by parts inequality requires an intricate cut-off and mollification procedure and the application of Hardy's inequality. One of the terms that have to be controlled during the proof is of the type

$$\frac{1}{\sigma}\int_0^T \int_{E^t\setminus E^{t,\sigma}} |u|^2 \,\mathrm{d}x\,\mathrm{d}t$$

for $0 < \sigma \ll 1$, where $E^{t,\sigma} := \{x \in E^t : \mathrm{dist}(x, \partial E^t) > \sigma\}$ denotes the inner parallel set of E^t. This is the point where we need to require the lower bound $p > \frac{2(n+1)}{n+2}$ on the growth exponent, since only in this case, a sufficiently strong Hardy–Sobolev inequality holds true that ensures that the above integrals vanish in the limit $\sigma \downarrow 0$.

The remaining part of the argument is divided into the proof of forward continuity and backward continuity in time. For the first part, we observe that in the interior of the domain, we have continuity in time since we can apply the standard embedding on small cylinders compactly contained in E. It remains to show that there is no loss of L^2-norm at the boundary in the limit $\tau \uparrow \tau_o$, which could happen if the domain shrank too fast. However, this loss of L^2-norm can be excluded by the inequality

$$\langle \partial_t u, \chi_{\Omega\times(\tau,\tau_o)} u\rangle \le \tfrac{1}{2}\|u(\tau_o)\|^2_{L^2(\Omega)} - \tfrac{1}{2}\|u(\tau)\|^2_{L^2(\Omega)} \tag{4.5}$$

for a.e. $\tau, \tau_o \in (0, T)$ with $\tau < \tau_o$, which is a consequence of (4.4).

The proof of the backward continuity exploits the fact that initial values are admitted in the L^2-sense. Hence, the crucial step in the proof is to show that the variational inequality can be localized to subdomains $E \cap \Omega \times (\tau_o, \tau)$ for every $\tau_o, \tau \in (0, T)$ with $\tau_o < \tau$. For the proof of this localization property, we again rely on the integration by parts formula (4.4). Moreover, the backward continuity in time is needed at this stage, since otherwise, the localization would only be possible for a.e. $\tau_o < \tau$.

4.3 Uniqueness of Variational Solutions

For a uniqueness result on possibly decreasing domains, we have to strengthen the one-sided growth condition for the complementary excess (4.2) to a two-sided condition, which guarantees that the domain neither decreases nor increases too fast in time. Recall that in [17] such a condition is already needed to prove regularity in the space $C^0([0, T); W^{1,2}(\mathbb{R}^n))$. For our uniqueness result, we assume

$$\mathbf{d}^c_H(E^s, E^t) \le M(t - s) \quad \text{for all } 0 \le s \le t < T, \tag{4.6}$$

with the *complementary Hausdorff distance*

$$\mathbf{d}^c_H(E^s, E^t) := \max\{\mathbf{e}^c(E^s, E^t), \mathbf{e}^c(E^t, E^s)\}.$$

Then, we obtained in [8, Theorem 2.9] the following uniqueness result.

Theorem 4.3 *Consider a variational integrand $f : \Omega \times \mathbb{R}^N \times \mathbb{R}^{Nn} \to [0, \infty]$ with the properties (2.3)$_1$ and (4.3), for some integrability exponent $p > \frac{2(n+1)}{n+2}$. For the non-cylindrical domain E we assume (4.1) and (4.6). Then, there exists at most one variational solution in the sense of Definition 2.1 that satisfies $\partial_t u \in (V^p(E))'$.*

Note that under the assumptions of Theorem 4.3 the solution constructed in Theorem 4.1 satisfies $\partial_t u \in (V^p(E))'$; see [8, Theorem 2.7]. With this respect, Theorem 4.3 ensures that our construction leads to a unique solution satisfying $\partial_t u \in (V^p(E))'$. However, it is not clear whether or not there are other variational solutions that do not possess this property.

For the proof of uniqueness, we again rely on the integration by parts formula (4.4). At this stage, we use it to recast the variational inequality (2.5) into the form

$$\iint_{E\cap\Omega_\tau} f(x,u,Du)\mathrm{d}x\mathrm{d}t \le \langle \partial_t u, \chi_{\Omega_\tau}(v-u)\rangle + \iint_{E\cap\Omega_\tau} f(x,v,Dv)\mathrm{d}x\mathrm{d}t \tag{4.7}$$

for every $v \in V_2^p(E)$ and every $\tau \in (0,T)$. Since in this form, the comparison map is not required to possess a weak time derivative in $L^2(\Omega_T,\mathbb{R}^N)$, it is possible to use another solution as comparison map in this inequality. Hence, given two solutions u_1 and u_2 of the Cauchy-Dirichlet problem with $\partial_t u_i \in (V^p(E))'$ for $i = 1,2$, a standard comparison procedure yields the estimate

$$\langle \partial_t(u_1-u_2), \chi_{\Omega_\tau}(u_1-u_2)\rangle \le 0$$

for every $\tau \in (0,T)$. Since $u_1 - u_2$ vanishes at the initial time $\tau = 0$, this yields the desired uniqueness if we have equality in the integration by parts formula (4.5) for $u_1 - u_2$ in place of u. However, this identity holds under the two-sided condition (4.6) on the non-cylindrical domain. This is the only point in the proof in which we also need to exclude domains that grow too fast in time.

4.4 Weak Solutions of the Associated Differential Equation

In the case of a differentiable integrand, we are able to show under the assumptions of Theorem 4.3 that every variational solution is a weak solution of the gradient flow associated to f.

Corollary 4.4 *Suppose that $f\colon \Omega\times\mathbb{R}^N\times\mathbb{R}^{Nn} \to [0,\infty]$ is a variational integrand with (2.3)$_1$ and (4.3) for which the partial maps $(u,\xi) \mapsto f(x,u,\xi)$ are of class C^1 for a.e. $x \in \Omega$. For the non-cylindrical domain E, we assume (4.1) and (4.2). Then every variational solution u of (2.5) with $\partial_t u \in (V^p(E))'$ is a weak solution of the differential equation*

$$\partial_t u - \operatorname{div}\big(D_\xi f(x,u,Du)\big) = -D_u f(x,u,Du) \quad \text{in } E.$$

The weak form of the differential equation is obtained from the strong formulation of the variational inequality (4.7) by the use of comparison maps of the form $v = u + s\varphi$ with $\varphi \in C_0^\infty(E,\mathbb{R}^N)$ and $s > 0$. Passing to the limit $s \downarrow 0$ then yields the weak formulation of the gradient flow associated to the integrand f.

References

1. P. Acquistapace, B. Terreni, A unified approach to abstract linear non autonomous parabolic equations. Rend. Sem. Mat. Univ. Padova **78**, 47–107 (1987)
2. B. Avelin, On time dependent domains for the degenerate p-parabolic equation: Carleson estimate and Hölder continuity. Math. Ann. **364**(1–2), 667–686 (2016)
3. C. Baiocchi, Regolarità e unicità della soluzione di una equazione differenziale astratta. Rend. Sem. Mat. Univ. Padova **35**, 380–417 (1965)
4. A. Björn, J. Björn, U. Gianazza, M. Parviainen, Boundary regularity for degenerate and singular parabolic equations. Calc. Var. Partial Differ. Equations **52**(3–4), 797–827 (2015)
5. A. Björn, J. Björn, U. Gianazza, The Petrovskiĭ criterion and barriers for degenerate and singular p-parabolic equations. Math. Ann. **368**(3–4), 885–904 (2017)
6. V. Bögelein, F. Duzaar, P. Marcellini, Existence of evolutionary variational solutions via the calculus of variations. J. Differ. Equations **256**(12), 3912–3942 (2014)
7. V. Bögelein, F. Duzaar, C. Scheven, P. Marcellini, Doubly nonlinear equations of porous medium type. Arch. Ration. Mech. Anal. **229**(2), 503–545 (2018)
8. V. Bögelein, F. Duzaar, C. Scheven, T. Singer, Existence of variational solutions in non-cylindrical domains. SIAM J. Math. Anal. **50**(3), 3007–3057 (2018)
9. S. Bonaccorsi, G. Guatteri, A variational approach to evolution problems with variable domains. J. Differ. Equations **175**(1), 51–70 (2001)
10. R. Brown, W. Hu, G. Lieberman, Weak solutions of parabolic equations in non-cylindrical domains. Proc. Am. Math. Soc. **125**(6), 1785–1792 (1997)
11. S. Byun, L. Wang, Parabolic equations in time dependent Reifenberg domains. Adv. Math. **212**(2), 797–818 (2007)
12. J. Calvo, M. Novaga, G. Orlandi, Parabolic equations on time dependent domains. J. Evol. Equ. **17**(2), 781–804 (2017)
13. P. Cannarsa, G. Da Prato, J. Zolésio, Evolution equations in noncylindrical domains. Atti Accad. Naz. Lincei Rend. Cl. Sci. Fis. Mat. Natur. (8) **83**, 73–77 (1989/1990)
14. E. Crampin, E. Gaffney, P. Maini, Reaction and diffusion on growing domains: scenarios for robust pattern formation. Bull. Math. Biol. **61**, 1093–1120 (1999)
15. E. De Giorgi, New problems on minimizing movements. Boundary value problems for PDES and applications. RMA Res. Notes Appl. Math. **29**, 81–98 (1993)
16. E. Fabes, N. Garofalo, S. Salsa, Comparison theorems for temperatures in noncylindrical domains. Atti Accad. Naz. Lincei Rend. Cl. Sci. Fis. Mat. Natur. (8) **77**(1–2), 1–12 (1984/1985)
17. U. Gianazza, G. Savaré, Abstract evolution equations on variable domains: an approach by minimizing movements. Ann. Scuola Norm. Sup. Pisa Cl. Sci. (4) **23**(1), 149–178 (1996)
18. S. Hofmann, J. Lewis, K. Nyström, Caloric measure in parabolic flat domains. Duke Math. J. **122**(2), 281–346 (2004)
19. T. Kilpeläinen, P. Lindqvist, On the Dirichlet boundary value problem for a degenerate parabolic equation. SIAM J. Math. Anal. **27**(3), 661–683 (1996)
20. E. Knobloch, R. Krechetnikov, Problems on time-varying domains: formulation, dynamics, and challenges. Acta Appl. Math. **137**, 123–157 (2015)
21. R. Krechetnikov, A linear stability theory on time-invariant and time-dependent spatial domains with symmetry: the drop splash problem. Dyn. Partial Differ. Equ. **8**, 47–67 (2011)
22. A. Lichnewsky, R. Temam, Pseudosolutions of the time-dependent minimal surface problem. J. Differ. Equations **30**(3), 340–364 (1978)
23. P. Lindqvist, A criterion of Petrowsky's kind for a degenerate quasilinear parabolic equation. Rev. Mat. Iberoamericana **11**(3), 569–578 (1995)
24. J. Lions, Sur les problèmes mixtes pour certains systèmes paraboliques dans des ouverts non cylindriques. Ann. Inst. Fourier **7**, 143–182 (1957)
25. G. Lumer, R. Schnaubelt, Time-dependent parabolic problems on non-cylindrical domains with inhomogeneous boundary conditions. J. Evol. Equ. **1**, 291–309 (2001)

26. J. Murray, *Mathematical biology. I. An introduction*. Interdisciplinary Applied Mathematics, vol. 17 (Springer, New York, 2002)
27. F. Paronetto, An existence result for evolution equations in non-cylindrical domains. NoDEA Nonlinear Differ. Equations Appl. **20**(6), 1723–1740 (2013)
28. G. Savaré, Parabolic problems with mixed variable lateral conditions: an abstract approach. J. Math. Pures Appl. (9) **76**(4), 321–351 (1997)
29. M. Shelley, F. Tiany, K. Wlodarski, Hele-Shaw flow and pattern formation in a time-dependent gap. Nonlinearity **10**, 1471–1495 (1997)
30. J. Stefan, Über die Theorie der Eisbildung, insbesondere über die Eisbildung im Polarmeere. Ann. Phys. Chem. **42**, 269–286 (1891)

A Compactness Result for the Sobolev Embedding via Potential Theory

Filippo Camellini, Michela Eleuteri, and Sergio Polidoro

Devoted to Emmanuele Di Benedetto in occasion of his 70th birthday

Abstract In this note we give a proof of the Sobolev and Morrey embedding theorems based on the representation of functions in terms of the fundamental solution of suitable partial differential operators. We also prove the compactness of the Sobolev embedding. We first describe this method in the classical setting, where the fundamental solution of the Laplace equation is used, to recover the classical Sobolev and Morrey theorems. We next consider degenerate Kolmogorov equations. In this case, the fundamental solution is invariant with respect to a non-Euclidean translation group and the usual convolution is replaced by an operation that is defined in accordance with this geometry. We recover some known embedding results and we prove the compactness of the Sobolev embedding. We finally apply our regularity results to a kinetic equation.

Keywords Sobolev spaces · Sobolev embedding · Morrey embedding · Compactness · Fundamental solution · Kolmogorov equation

1 Introduction

Sobolev and Morrey embedding theorems are fundamental tools in the regularity theory for Elliptic and Parabolic second order Partial Differential Equations (*PDEs* in the sequel). In particular, they play a crucial role in the natural setting for the study of uniformly elliptic PDEs in divergence form, that is the Sobolev space $W^{1,p}$.

Investigation supported by I.N.d.A.M.

F. Camellini · M. Eleuteri · S. Polidoro (✉)
Dipartimento di Scienze Fisiche Informatiche e Matematiche, Modena, Italy
e-mail: michela.eleuteri@unimore.it; sergio.polidoro@unimore.it

V. Vespri et al. (eds.), *Harnack Inequalities and Nonlinear Operators*, Springer INdAM Series 46, https://doi.org/10.1007/978-3-030-73778-8_4

There are several proofs of the Sobolev and Morrey embedding theorems, all of them rely on some integral representation of a general function $u \in W^{1,p}$ in terms of its gradient. Here we focus in particular on representation formulas based on the fundamental solution of the Laplace equation.

Consider a function $u \in C_0^\infty(\mathbb{R}^n)$. By the very definition of fundamental solution Γ, the following identity holds

$$u(x) = -\int_{\mathbb{R}^n} \Gamma(x-y)\Delta u(y)\, dy, \qquad \text{for every } x \in \mathbb{R}^n, \tag{1.1}$$

and an integration by parts immediately gives

$$u(x) = \int_{\mathbb{R}^n} \big\langle \nabla_y \Gamma(x-y), \nabla u(y)\big\rangle dy, \qquad \text{for every } x \in \mathbb{R}^n, \tag{1.2}$$

where $\langle\cdot,\cdot\rangle$ and ∇ denote the usual inner product in $\mathbb{R}^n$ and the gradient, respectively. We recall that the gradient of the fundamental solution of the Laplace equation writes as follows

$$\nabla\Gamma(x-y) = -\frac{1}{n\omega_n|x-y|^n}(x-y), \qquad x \neq y, \tag{1.3}$$

where ω_n is the measure of the n-dimensional unit ball. In particular, $\nabla\Gamma$ is an homogeneous function of degree $-n+1$, and there exists a positive constant c_n such that

$$|\nabla\Gamma(x-y)| \le c_n|x-y|^{1-n}, \tag{1.4}$$

thus (1.2) yields the following inequality:

$$|u(x)| \le c_n \int_{\mathbb{R}^n} |x-y|^{1-n}|\nabla u(y)| dy. \tag{1.5}$$

The Young inequality for convolution with homogeneous kernels (see, for instance, Theorem 1, p. 119 in [11]) then gives

$$\|\nabla\Gamma * \nabla u\|_{L^{p^*}(\mathbb{R}^n)} \le C_p\, \|\nabla u\|_{L^p(\mathbb{R}^n)}, \qquad 1 < p < n, \tag{1.6}$$

where $p^* = \frac{pn}{n-p}$ is the Sobolev conjugate of p, and C_p is a positive constant which only depends on p and on the dimension n. Here and in the sequel the dependence on n will be often omitted. As a consequence we find

$$\|u\|_{L^{p^*}(\mathbb{R}^n)} \le C_p\, \|\nabla u\|_{L^p(\mathbb{R}^n)}, \qquad \text{for every } u \in C_0^\infty(\mathbb{R}^n), \quad 1 < p < n. \tag{1.7}$$

From the above inequality we plainly obtain the following Sobolev inequality for any open set $\Omega \subseteq \mathbb{R}^n$

$$\|u\|_{L^q(\Omega)} \leq C_{p,q} \|u\|_{W^{1,p}(\Omega)}, \qquad \text{for every } u \in W_0^{1,p}(\Omega), \tag{1.8}$$

with $1 < p < n$ and $p \leq q \leq p^*$. Here $C_{p,q}$ is a positive constant which only depends on p, q and n. By a standard argument (1.6) also gives the Sobolev embedding theorem for $W^{1,p}(\Omega)$ provided that the boundary of Ω is sufficiently smooth.

The Morrey inequality (see Theorem 2.4 below) can be obtained by the representation formula (1.2), by using the following fact: there exists a positive constant M_n, only depending on n, such that with

$$|\partial_{x_j}\Gamma(x) - \partial_{x_j}\Gamma(y)| \leq M_n \frac{|x-y|}{|x|^n}, \qquad \text{for } j = 1, \ldots, n, \tag{1.9}$$

for every $x, y \in \mathbb{R}^n \setminus \{0\}$ such that $|x-y| \leq |x|/2$. Indeed, a rather simple argument based on (1.9) provides us with the following bound: if $u \in W_0^{1,p}(\Omega)$, with $p > n$, then

$$|u(x) - u(y)| \leq \widetilde{C}_p \|\nabla u\|_{L^p(\Omega)} |x-y|^{1-\frac{n}{p}}, \qquad \text{for every } x, y \in \Omega, \tag{1.10}$$

for some positive constant $\widetilde{C}_p$ only depending on p and n.

It is worth noting that the inequality (1.9) can be also used to prove the compactness of the Sobolev embedding (1.8) for $p < q < p^*$, if Ω is a bounded open set. As we will see in the sequel, the following estimates holds for $p < q < p^*$: there exists a positive constant $\widetilde{C}_{p,q}$ such that

$$\|u(h + \cdot) - u\|_{L^q(\Omega)} \leq \widetilde{C}_{p,q} \|\nabla u\|_{L^p(\Omega)} |h|^{n\left(\frac{1}{q} - \frac{1}{p^*}\right)}, \tag{1.11}$$

for every $u \in C_0^\infty(\Omega)$ and for every $h \in \mathbb{R}^n$ sufficiently small. Note that the exponent in the right hand side of (1.11) belongs to the interval $]0, 1[$ if, and only if, $p < q < p^*$, then in this case we have

$$\|u(h + \cdot) - u\|_{L^q(\Omega)} \to 0 \qquad \text{as} \qquad |h| \to 0.$$

This inequality provides us with the *integral uniform continuity*, which is needed for the compactness in the L^q spaces. We also observe that $n\left(\frac{1}{q} - \frac{1}{p^*}\right) \to 1$ as $q \to p$. We then retrieve a known result contained for instance in [11, Chapter V, Section 3.5].

The advantage of the method described above, with respect to other ones, is in that it only requires the existence of a *fundamental solution* and its *homogeneity* properties. In particular, it applies to the function spaces introduced by Folland [4] for the study second order linear differential operators that satisfy the Hörmander's condition (see [6]). It should be noticed that this approach has also a drawback, in that it does not provide us with the Sobolev inequality for $p = 1$. On the other hand it is unifying, as it gives the Sobolev and Morrey embedding theorems and a compactness result by using a single representation formula.

To clarify the use of this method to the study of the so-called Hörmander's operators we next focus on the degenerate Kolmogorov $\mathscr{L}_0$ on $\mathbb{R}^{2n+1}$, which is one of the simplest examples belonging to this class. Let Ω be an open subset of $\mathbb{R}^{2n+1}$ and let u be a smooth real valued function defined on Ω. We denote the variable of $\mathbb{R}^{2n+1}$ as follows $z = (x, y, t) \in \mathbb{R}^n \times \mathbb{R}^n \times \mathbb{R}$, and we set

$$\mathscr{L}_0 u := \Delta_x u + \langle x, \nabla_y u\rangle - \partial_t u, \qquad \Delta_x u := \sum_{j=1}^{n} \partial^2_{x_j} u. \tag{1.12}$$

As we will see in the sequel (see Eq. (4.1) below) the function Γ defined as

$$\begin{cases} \Gamma(x, y, t) = \frac{\widetilde{c}_n}{t^{2n}} \exp\left(-\frac{|x|^2}{t} - 3\frac{\langle x, y\rangle}{t^2} - 3\frac{|y|^2}{t^3}\right), & \text{for } (x, y, t) \in \mathbb{R}^{2n} \times]0, +\infty[, \\ \Gamma(x, y, t) = 0, & \text{for } (x, y, t) \in \mathbb{R}^{2n} \times] - \infty, 0], \end{cases}$$

is the fundamental solution of $\mathscr{L}_0$. Here $\widetilde{c}_n = \frac{3^{n/2}}{(2\pi)^n}$. In particular, in analogy with the heat equation, we have that the function u defined as

$$\begin{aligned} u(x, y, t) = &\int_{\mathbb{R}^{2n}} \Gamma(x - \xi, y + t\xi - \eta, t - t_0)\varphi(\xi, \eta) d\xi\, d\eta - \\ &\int_{\mathbb{R}^{2n} \times]t_0, t[} \Gamma(x - \xi, y + (t - \tau)\xi - \eta, t - \tau) f(\xi, \eta, \tau) d\xi\, d\eta\, d\tau \end{aligned} \tag{1.13}$$

is a solution to the following Cauchy problem

$$\begin{cases} \mathscr{L}_0 u = f & \text{in } \mathbb{R}^{2n} \times]t_0, +\infty[, \\ u_{|t=t_0} = \varphi & \text{in } \mathbb{R}^{2n}. \end{cases}$$

whenever f and φ are bounded continuous functions.

A remarkable fact is that a kind of convolution is hidden in the expression (1.13). More specifically, we define the operation "$\circ$" by setting

$$(x, y, t) \circ (\xi, \eta, \tau) := (x + \xi, y + \eta + \tau x, t + \tau), \qquad (x, y, t), (\xi, \eta, \tau) \in \mathbb{R}^{2n+1}, \tag{1.14}$$

and we note that $\left(\mathbb{R}^{2n+1}, \circ\right)$ is a non commutative group. The identity of the group is $(0, 0, 0)$ and the inverse of (x, y, t) is $(-x, -y + xt, -t)$. With this notation, it is easy to check that the expression appearing in (1.13) can be written as follows

$$(x - \xi, y + (t - \tau)\xi - \eta, t - \tau) = (\xi, \eta, \tau)^{-1} \circ (x, y, t).$$

Moreover, the group $\left(\mathbb{R}^{2n+1}, \circ\right)$ is homogeneous with respect to the dilation defined as $d_r(x, y, t) := \left(rx, r^3 y, r^2 t\right)$, in the sense that

$$d_r\big((x, y, t) \circ (\xi, \eta, \tau)\big) = d_r(x, y, t) \circ d_r(\xi, \eta, \tau), \qquad (x, y, t), (\xi, \eta, \tau) \in \mathbb{R}^{2n+1}, r > 0. \tag{1.15}$$

This algebraic structure was introduced and studied by Lanconelli and Polidoro in [7]. In [7] it was also noticed that Γ is homogeneous of degree $-4n$ with respect to $(d_t)_{r>0}$, that is

$$\Gamma\big(d_r(x, y, t)\big) = \frac{1}{r^{4n}}\Gamma(x, y, t), \qquad (x, y, t) \in \mathbb{R}^{2n+1}, r > 0. \tag{1.16}$$

Moreover, if we let $z = (x, y, t)$, $\zeta = (\xi, \eta, \tau)$, then (1.13) can be written as follows

$$u(z) = \int_{\mathbb{R}^{2n}} \Gamma((\xi, \eta, t_0)^{-1} \circ z)\varphi(\xi, \eta) d\xi\, d\eta - \int_{\mathbb{R}^{2n} \times]t_0, t[} \Gamma(\zeta^{-1} \circ z) f(\zeta) d\zeta. \tag{1.17}$$

In particular, if $u \in C_0^\infty(\mathbb{R}^{2n+1})$ and $\mathrm{supp}(u) \subset \left\{t > t_0\right\}$, then we have that

$$u(z) = -\int_{\mathbb{R}^{2n+1}} \Gamma(\zeta^{-1} \circ z)\mathscr{L}_0 u(\zeta)\, d\zeta, \qquad \text{for every } z \in \mathbb{R}^{2n+1}, \tag{1.18}$$

which is analogous to (1.1). Summarizing: the operation in (1.18) is considered here as a *convolution* with respect to the non-Euclidean operation "$\circ$" defined in (1.14), with a kernel Γ that is homogeneous whit respect to the *anisotropic* dilation d_r. Based on this representation formula, we prove Sobolev and Morrey theorems for solutions to Kolmogorov equations in divergence form $\mathscr{L}u = \mathrm{div}_x F + f$, where

$$\mathscr{L}u := \mathrm{div}_x\, (A(z)\nabla_x u) + \langle x, \nabla_y u\rangle - \partial_t u. \tag{1.19}$$

Here A is a $n \times n$ symmetric matrix with bounded and measurable coefficients and, for every vector field $F \in C^1(\mathbb{R}^{2n+1}, \mathbb{R}^n)$ we denote $\mathrm{div}_x F(x, y, t) := \sum_{j=1}^n \partial_{x_j} F_j(x, y, t)$. In order to simplify our treatment, we suppose that $F = 0$ and $f = 0$, so that u is a solution of $\mathscr{L}u = 0$. In this case we have that $\mathscr{L}_0 u = \mathrm{div}_x (I_n - A)\nabla_x u$, where I_n denotes the $n \times n$ identity matrix. Then, an integration by parts in (1.18) gives

$$u(z) = \int_{\mathbb{R}^{2n+1}} \langle (I_n - A(\zeta))\nabla_\xi \Gamma(\zeta^{-1} \circ z), \nabla_\xi u(\zeta)\rangle\, d\zeta, \tag{1.20}$$

for every solution u to $\mathscr{L}u = 0$. It is known that the derivatives $\partial_{\xi_1}\Gamma, \dots, \partial_{\xi_n}\Gamma$ are homogeneous functions of degree $-(2n+1)$ with respect to the dilation $(d_r)_{r>0}$. Moreover, the coefficients of the matrix $I_n - A$ are bounded, then the above identity provides us with the analogous of (1.2) for the solutions u to the equation $\mathscr{L}u = 0$.

We point out that only the derivatives with respect to the first n variables of the gradient of u appear in the representation formula (1.20), then a Sobolev inequality holding for all functions cannot be obtained from (1.20), because of the lack of information on the remaining n direction. Nevertheless, this formula is used by Cinti, Pascucci, and Polidoro in [3, 9] to prove a Sobolev embedding theorem for *solutions* to the Kolmogorov equation $\mathscr{L}u = 0$. Indeed, in [3, 9] the Sobolev theorem for *solutions* is combined with a Caccioppoli inequality, still for *solutions*, in order to apply the Moser's iterative method and prove an L^∞_{loc} estimate for the solutions to $\mathscr{L}u = 0$. We also recall that a Morrey result for the solutions to $\mathscr{L}u = \text{div}_x F$ was proven by Manfredini and Polidoro in [8], and later by Polidoro and Ragusa in [10] by the same method used here.

In this note we are concerned with the compactness of the Sobolev embedding for the solutions to $\mathscr{L}u = 0$ for a family of degenerate Kolmogorov equations, defined on $\mathbb{R}^{N+1}$, that will be still denoted by $\mathscr{L}$. As we will see in Sect. 3, the operator (1.19) is the prototype of this family of degenerate operators, and in this case, $N = 2n$. In Sect. 3 we introduce the notation that will be used in the following part of this introduction, and we will state the conditions (H.1) and (H.2) that ensure that the *principal part* $\mathscr{L}_0$ of $\mathscr{L}$ has a smooth fundamental solution Γ, which is invariant with respect to a translation analogous to (1.14), and homogeneous of degree $-Q$, with respect to a dilation analogous to (1.15). We will refer to the positive integer $Q+2$ as *homogeneous dimension* of the space $\mathbb{R}^{N+1}$ and plays the role of n in the Euclidean setting $\mathbb{R}^n$ where the elliptic operators are studied. In the sequel p^* and p^{**} denote the positive numbers such that

$$\frac{1}{p^*} = \frac{1}{p} - \frac{1}{Q+2}, \qquad \frac{1}{p^{**}} = \frac{1}{p} - \frac{2}{Q+2}. \tag{1.21}$$

Clearly, p^* and p^{**} are finite and positive whenever $1 \le p < Q+2$ and $1 \le p < \frac{Q+2}{2}$, respectively.

Our main result is the following Theorem. It provides us with some estimates of the convolution of a function belonging to some L^p space with the fundamental solution Γ and with its derivatives $\partial_{x_j}\Gamma$, $j = 1, \dots, m_0$, with $m_0 \le N$. These estimates, applied to the representation formula for solutions to $\mathscr{L}u = 0$ given in Theorem 4.1, yield Sobolev theorems, Morrey theorems and the compactness of the Sobolev embedding.

Theorem 1.1 *Let $\mathscr{L}$ be an operator in the form* (3.1)*, satisfying the hypotheses (H.1) and (H.2) in Sect.* 3*, and let Γ be the fundamental solution of its* principal part. *Let also $Q+2$ be the homogeneous dimension of the space $\mathbb{R}^{N+1}$, and let p*

be such that $1 \le p < +\infty$. *For every* $f, g_j \in L^p(\mathbb{R}^{N+1})$ *we let* u, v_j *be defined as follows*

$$u(z) = \int_{\mathbb{R}^{N+1}} \Gamma(\zeta^{-1} \circ z) f(\zeta)\, \mathrm{d}\zeta, \quad v_j(z) = \int_{\mathbb{R}^{N+1}} \partial_{x_j} \Gamma(\zeta^{-1} \circ z) g_j(\zeta)\, \mathrm{d}\zeta, \quad j = 1, \dots, m_0.$$

Then, for every $j = 1, \dots, m_0$ *we have:*

- *(Sobolev) if* $1 < p < Q+2$, *then there exists a positive constant* C_p *such that*

$$\|v_j\|_{L^{p^*}(\mathbb{R}^{N+1})} \le C_p \, \|g_j\|_{L^p(\mathbb{R}^{N+1})},$$

- *(Compactness) if moreover* $p < q < p^*$, *then there exists a positive constant* $\widetilde{C}_{p,q}$ *such that*

$$\|v_j(\cdot \circ h) - v_j\|_{L^q(\mathbb{R}^{N+1})} \le \widetilde{C}_{p,q} \, \|g_j\|_{L^p(\mathbb{R}^{N+1})} \, \|h\|^{(Q+2)\left(\frac{1}{q} - \frac{1}{p^*}\right)},$$

  for every $h \in \mathbb{R}^{N+1}$,
- *(Morrey) if* $p > Q+2$, *then there exists a positive constant* $\widetilde{C}_p$ *such that*

$$|v_j(z) - v_j(\zeta)| \le \widetilde{C}_p \, \|g_j\|_{L^p(\mathbb{R}^{N+1})} \|\zeta^{-1} \circ z\|^{1 - \frac{Q+2}{p}}, \quad \text{for every } z, \zeta \in \mathbb{R}^{N+1}.$$

We also have

- *(Sobolev) if* $1 < p < \dfrac{Q+2}{2}$, *then there exists a positive constant* C_p *such that*

$$\|u\|_{L^{p^{**}}(\mathbb{R}^{N+1})} \le C_p \, \|f\|_{L^p(\mathbb{R}^{N+1})},$$

- *(Compactness) if* $p^* < q < p^{**}$, *then there exists a positive constant* $\widetilde{C}_{p,q}$ *such that*

$$\|u(\cdot \circ h) - u\|_{L^q(\mathbb{R}^{N+1})} \le \widetilde{C}_{p,q} \, \|f\|_{L^p(\mathbb{R}^{N+1})} \, \|h\|^{(Q+2)\left(\frac{1}{q} - \frac{1}{p^{**}}\right)},$$

  for every $h \in \mathbb{R}^{N+1}$,
- *(Morrey) if* $\dfrac{Q+2}{2} < p < Q+2$, *then there exists a positive constant* $\widetilde{C}_p$ *such that*

$$|u(z) - u(\zeta)| \le \widetilde{C}_p \, \|f\|_{L^p(\mathbb{R}^{N+1})} \|\zeta^{-1} \circ z\|^{2 - \frac{Q+2}{p}}, \quad \text{for every } z, \zeta \in \mathbb{R}^{N+1}.$$

From the above result and a representation formula for the solution to $\mathscr{L} u = 0$ we obtain the following result.

Theorem 1.2 *Let Ω be an open set of $\mathbb{R}^{N+1}$, and let u be a weak solution to $\mathscr{L}u = 0$ in Ω. Suppose that $u, \partial_{x_1}u, \ldots, \partial_{x_{m_0}}u \in L^p(\Omega)$. Then for every compact set $K \subset \Omega$, there exist a positive constant $\widetilde{\varrho}$ such that we have:*

- *(Sobolev embedding) if $1 < p < Q+2$, then there exists a positive constant C_p such that*

$$\|u\|_{L^{p^*}(K)} \le C_p\left(\|u\|_{L^p(\Omega)} + \sum_{j=1}^{m_0}\|\partial_{x_j}u\|_{L^p(\Omega)}\right),$$

- *(Compactness) if moreover $p < q < p^*$, then there exists a positive constant $\widetilde{C}_{p,q}$ such that*

$$\|u(\cdot\circ h) - u\|_{L^q(K)} \le \widetilde{C}_{p,q}\left(\|u\|_{L^p(\Omega)} + \sum_{j=1}^{m_0}\|\partial_{x_j}u\|_{L^p(\Omega)}\right)\|h\|^{(Q+2)\left(\frac{1}{q}-\frac{1}{p^*}\right)},$$

  for every $h \in \mathbb{R}^{N+1}$ such that $\|h\| \le \widetilde{\varrho}$,
- *(Morrey embedding) if $p > Q+2$, then there exists a positive constant $\widetilde{C}_p$ such that*

$$|u(z) - u(\zeta)| \le \widetilde{C}_p\left(\|u\|_{L^p(\Omega)} + \sum_{j=1}^{m_0}\|\partial_{x_j}u\|_{L^p(\Omega)}\right)\|\zeta^{-1}\circ z\|^{1-\frac{Q+2}{p}},$$

  for every $z, \zeta \in K$ such that $\|\zeta^{-1}\circ z\| \le \widetilde{\varrho}$.

The following theorem is related to the main result of the article [1] by Bouchut, where the regularity of the solution of the kinetic equation

$$\partial_t f + \langle v, \nabla_x f\rangle = g, \qquad (t, x, v) \in \Omega \subseteq \mathbb{R}\times\mathbb{R}^n\times\mathbb{R}^n, \tag{1.22}$$

is considered. Note that the differential operator appearing in the left hand side of (1.22) agrees with the first order part of $\mathscr{L}$ defined in (1.19). Actually, the notation of the following result refers to this operator, and, in particular, the homogeneous dimension of the space $\mathbb{R}^{2n+1}$ is in this case $Q + 2 = 4n + 2$.

Theorem 1.3 *Let Ω be an open set of $\mathbb{R}^{2n+1}$, and let $f \in L^2_{\text{loc}}(\Omega)$ be a weak solution to (1.22). Suppose that $g, f, \partial_{v_1}f, \ldots, \partial_{v_n}f \in L^p(\Omega)$. Then for every compact set $K \subset \Omega$, there exist a positive constant $\widetilde{\varrho}$ such that we have:*

- *(Sobolev embedding) if $1 < p < 4n+2$, then there exists a positive constant C_p such that*

$$\|f\|_{L^{p^*}(K)} \le C_p\left(\|g\|_{L^p(\Omega)} + \|f\|_{L^p(\Omega)} + \sum_{j=1}^{n}\|\partial_{v_j}f\|_{L^p(\Omega)}\right),$$

- *(Compactness) if moreover $p < q < p^*$, then there exists a positive constant $\widetilde{C}_{p,q}$ such that*

$$\|f(\cdot \circ h) - f\|_{L^q(K)} \le \widetilde{C}_{p,q}\left(\|g\|_{L^p(\Omega)} + \|f\|_{L^p(\Omega)} + \sum_{j=1}^{n}\|\partial_{v_j} f\|_{L^p(\Omega)}\right)\|h\|^{(4n+2)\left(\frac{1}{q}-\frac{1}{p^*}\right)},$$

for every $h \in \mathbb{R}^{2n+1}$ such that $\|h\| \le \widetilde{\varrho}$,
- *(Morrey embedding) if $p > 4n+2$, then there exists a positive constant $\widetilde{C}_p$ such that*

$$|f(z) - f(\zeta)| \le \widetilde{C}_p\left(\|g\|_{L^p(\Omega)} + \|f\|_{L^p(\Omega)} + \sum_{j=1}^{n}\|\partial_{v_j} f\|_{L^p(\Omega)}\right)\|\zeta^{-1} \circ z\|^{1-\frac{4n+2}{p}},$$

for every $z, \zeta \in K$ such that $\|\zeta^{-1} \circ z\| \le \widetilde{\varrho}$.

The proof of Theorems 1.2 and 1.3 is given in Sect. 4.

We next give some comments to our main results. We still refer here to the notation relevant to the operator $\mathscr{L}$ defined in (1.19), and to the representation formula (1.20). As we said above, it holds for solutions to $\mathscr{L}u = 0$ then, for this reason, it seems to be weaker than the usual Sobolev inequality. On the other hand, due to the strong degeneracy of the operator $\mathscr{L}$, its natural Sobolev space $W^{1,p}_{\mathscr{L}}$ is the space of the functions $u \in L^p$ with weak derivatives $\partial_{x_1}u, \dots, \partial_{x_n}u \in L^p$. In particular, it is impossible to prove a Sobolev inequality unless some information is given on u with respect to the remaining variables $y_1, \dots, y_n$ and t. We obtain this missing information from the fact that u is a solution to $\mathscr{L}u = 0$ (or, in a more general case, to $\mathscr{L}u = \mathrm{div}_x F + f$). We also note that the regularity property of the operator $\mathscr{L}$ is quite unstable. Indeed, let us fix any $x_0 \in \mathbb{R}^n$ and consider the operator $\widetilde{\mathscr{L}}_0$, acting on $(x, y, t) \in \mathbb{R}^{2n+1}$ as follows

$$\widetilde{\mathscr{L}}_0 u := \Delta_x u + \langle x_0, \nabla_y u\rangle - \partial_t u.$$

Its natural Sobolev spaces agrees with that of $\mathscr{L}$, however it is known that a fundamental solution for $\widetilde{\mathscr{L}}_0$ does not exists and our method for the proof of the Sobolev inequality fails in this case. Actually, it is not difficult to check that the Sobolev inequality does not hold for the solutions to $\widetilde{\mathscr{L}}_0 u = 0$.

We conclude this discussion with a simple remark. Also when we consider the more familiar uniformly parabolic equations, we find that the natural Sobolev space only contains the *spatial* derivatives, and it is not possible to find a simple natural space for the time derivative. As a matter of facts, several regularity results for parabolic equations depend on some *fractional Sobolev spaces*. The situation becomes more complicated when we consider second order PDEs with non-negative characteristic form analogous to $\mathscr{L}$. An alternative approach to our method, that only relies on a representation formula in terms of the fundamental solution, is the use of *fractional Sobolev spaces* (we refer to the articles by Bochut [1], see also

Golse, Imbert, Mouhot, and Vasseur [5]) to recover the missing information with respect to the variables $y_1, \ldots, y_n$ and t.

This article is organized as follows. In Sect. 2 we give a comprehensive proof of the Sobolev embedding, of its compactness, and the Morrey embedding, following the method above outlined. In Sect. 3 we recall the tools of the Real Analysis on Lie groups we need to prove Theorem 1.1, and we give its proof. In Sect. 4 we discuss some applications of Theorem 1.1 to the solutions of $\mathscr{L}u = 0$. Section 5 contains some comments about the possible extension of Theorem 1.1 to a family of more general operators considered by Cinti and Polidoro in [2].

2 Continuous and Compact Embeddings: The Euclidean Case

In this section we give a comprehensive proof of the Sobolev embedding (1.8), the Morrey embedding (1.10), and of the inequality (1.11) from which the compactness of the Sobolev embedding follows. As said in the Introduction, all these results rely on the representation formula (1.2), which gives the bound (1.5) that we recall below

$$|u(x)| \le c_n \int_{\mathbb{R}^n} |x-y|^{1-n} |\nabla u(y)|\, dy, \qquad \text{for every } u \in C_0^\infty(\mathbb{R}^n).$$

With this aim, we first recall the weak Young inequality that gives the Sobolev embedding, then we prove (1.9) and we deduce from this and (1.5) the Morrey embedding (1.10), and that stated in the inequality (1.11).

2.1 *Some Preliminary Results*

For a given positive α we denote by K_α any continuous homogeneous function of degree $-\alpha$, that is a function satisfying

$$K_\alpha(rx) = r^{-\alpha} K_\alpha(x), \quad \text{for every } x \in \mathbb{R}^n \setminus \{0\}, \text{ and } r > 0.$$

We easily see that

$$|K_\alpha(x)| \le \frac{c_\alpha}{|x|^\alpha}, \qquad \text{for every } x \in \mathbb{R}^n \setminus \{0\}, \tag{2.1}$$

where $c_\alpha := \max_{|x|=1} |K_\alpha(x)|$. In particular, K_α belongs to the space $L^q_{\text{weak}}(\mathbb{R}^n)$, for $q = \frac{n}{\alpha}$, that is

$$\text{meas}\big\{x \in \mathbb{R}^n \mid |K_\alpha(x)| \ge \lambda\big\} \le \left(\frac{C}{\lambda}\right)^q, \qquad \text{for every} \quad \lambda > 0, \tag{2.2}$$

for some non-negative constant C. Here meas E denotes the Lebesgue measure of the set E. Moreover we define the *seminorm* of K_α as follows

$$\|K_\alpha\|_{L^q_{\text{weak}}(\mathbb{R}^n)} := \inf\{C \geq 0 \mid (2.2) \text{ holds}\}.$$

From (2.1) it plainly follows that $C \leq c_\alpha \omega_n^{\alpha/n}$. We next recall two elementary inequalities that will be useful in the sequel. For every $R > 0$ we have that:

- $K_\alpha \in L^q(\{x \in \mathbb{R}^n \mid |x| \leq R\})$ if, and only if, $q < \dfrac{n}{\alpha}$. Moreover, there exists a positive constant $c_{\alpha,q}$, only depending on K_α, n and q, such that

$$\|K_\alpha\|_{L^q(\{x\in\mathbb{R}^n \mid |x|\leq R\})} \leq c_{\alpha,q} R^{\frac{n}{q}-\alpha}, \tag{2.3}$$

- $K_\alpha \in L^q(\{x \in \mathbb{R}^n \mid |x| \geq R\})$ if, and only if, $q > \dfrac{n}{\alpha}$. Moreover, there exists a positive constant $c_{\alpha,q}$, only depending on K_α, n and q, such that

$$\|K_\alpha\|_{L^q(\{x\in\mathbb{R}^n \mid |x|\geq R\})} \leq c_{\alpha,q} R^{\frac{n}{q}-\alpha}. \tag{2.4}$$

The following weak Young inequality holds (see Theorem 1, p. 119 in [11], where this result is referred to as Hardy-Littlewood-Sobolev theorem for fractional integration).

Theorem 2.1 *Let K_α be a continuous homogeneous function of degree $-\alpha$, with $0 < \alpha < n$. Let p, q be such that $1 \leq p < q < +\infty$ and that $1 + \dfrac{1}{q} = \dfrac{1}{p} + \dfrac{\alpha}{n}$. Then, for every $f \in L^p(\mathbb{R}^n)$ the integral $K_\alpha * f(x)$ is convergent for almost every $x \in \mathbb{R}^n$. Moreover,*

- *If $p > 1$, then there exists $C_{\alpha,p} > 0$ such that $\|K_\alpha * f\|_{L^q(\mathbb{R}^n)} \leq C_{\alpha,p}\|f\|_{L^p(\mathbb{R}^n)}$,*
- *If $p = 1$, then there exists $C_{\alpha,1} > 0$ such that $\|K_\alpha * f\|_{L^q_{\text{weak}}(\mathbb{R}^n)} \leq C_{\alpha,1}\|f\|_{L^1(\mathbb{R}^n)}$.*

In order to prove the Morrey embedding (1.10) and the compactness estimate (1.11), we state and prove the following lemma.

Lemma 2.2 *Let $K_\alpha \in C^1(\mathbb{R}^n\backslash\{0\})$ be any homogeneous function of degree $-\alpha$, with $0 < \alpha < n$. Then there exists a positive constant M_α such that*

$$|K_\alpha(x) - K_\alpha(y)| \leq M_\alpha \frac{|x-y|}{|x|^{\alpha+1}}, \quad \textit{for every} \quad x, y \in \mathbb{R}^n\backslash\{0\} \textit{ such that } |x-y| \leq \frac{|x|}{2}.$$

Proof We first prove the result for x such that $|x| = 1$. In this case $|x - y| \leq \dfrac{1}{2}$ and by the Mean Value Theorem there exists $\theta \in (0, 1)$ such that

$$|K_\alpha(x) - K_\alpha(y)| = |\langle (x-y), \nabla K_\alpha(\theta x + (1-\theta y))\rangle| \leq M_\alpha |x-y|,$$

where

$$M_\alpha = \max_{\frac{1}{2}\le|z|\le\frac{3}{2}} |\nabla K_\alpha(z)|. \tag{2.5}$$

Consider now a general choice of $x, y \in \mathbb{R}^n\setminus\{0\}$ with $|x-y| \le \frac{|x|}{2}$. Being K_α homogeneous of degree $-\alpha$, we obtain

$$|K_\alpha(x) - K_\alpha(y)| = \frac{1}{|x|^\alpha}\left|K_\alpha\left(\frac{x}{|x|}\right) - K_\alpha\left(\frac{y}{|x|}\right)\right| \le \frac{M_\alpha}{|x|^\alpha}\left|\frac{x}{|x|} - \frac{y}{|x|}\right| = M_\alpha \frac{|x-y|}{|x|^{\alpha+1}},$$

being M_α as in (2.5), because $\left|\frac{x}{|x|}\right| = 1$. □

In order to prove the Morrey embedding (1.10) and the compactness of the Sobolev embedding for $p < q < p^*$ we rely on the following argument. We choose any $u \in C_0^\infty(\mathbb{R}^n)$, $h \in \mathbb{R}^n$ and we set

$$v(x) := u(x+h) - u(x), \qquad \text{for every} \quad x \in \mathbb{R}^n, \tag{2.6}$$

then

$$\begin{aligned} v(x) = &\int_{\{y\in\mathbb{R}^n:|x+h-y|\ge 2|h|\}} \langle \nabla\Gamma(x+h-y) - \nabla\Gamma(x-y), \nabla u(y)\rangle\, dy \\ &+ \int_{\{y\in\mathbb{R}^n:|x+h-y|<2|h|\}} \langle \nabla\Gamma(x+h-y), \nabla u(y)\rangle\, dy \\ &+ \int_{\{y\in\mathbb{R}^n:|x+h-y|<2|h|\}} -\langle \nabla\Gamma(x-y), \nabla u(y)\rangle\, dy =: I_A(x) + I_B(x) + I_C(x). \end{aligned} \tag{2.7}$$

We next rely on Lemma 2.2 and on (1.3) to estimate the terms I_A, I_B and I_C as follows

$$\begin{aligned} |I_A(x)| &\le M|h| \int_{\{y\in\mathbb{R}^n:|x+h-y|\ge 2|h|\}} \frac{1}{|x+h-y|^n} |\nabla u(y)|\, dy, \\ |I_B(x)| &\le c_n \int_{\{y\in\mathbb{R}^n:|x+h-y|<2|h|\}} \frac{1}{|x+h-y|^{n-1}} |\nabla u(y)|\, dy, \\ |I_C(x)| &\le c_n \int_{\{y\in\mathbb{R}^n:|x+h-y|<2|h|\}} \frac{1}{|x-y|^{n-1}} |\nabla u(y)|\, dy, \end{aligned} \tag{2.8}$$

where $M := \max_{\frac{1}{2}\le|z|\le\frac{3}{2},\, j,k=1,\dots,n} |\partial^2_{x_j x_k}\Gamma(z)|$.

2.2 *The Sobolev and Morrey Embedding Theorems*

As we said in the Introduction, Theorem 2.1 combined with (1.5) immediately yields the following result

Theorem 2.3 *Let $1 < p < n$. Then there exists $C_p > 0$ such that:*

$$\|u\|_{p^*} \leq C_p \|\nabla u\|_p \quad \textit{for every} \quad u \in L^1_{\text{loc}}(\mathbb{R}^n) \quad \textit{such that} \quad \nabla u \in L^p(\mathbb{R}^n),$$

where $p^* = \dfrac{np}{n-p}$.

We next turn our attention on the Morrey's Theorem.

Theorem 2.4 (Morrey's Theorem) *Let $u : \mathbb{R}^n \to \mathbb{R}$ and $p > n$. If $\nabla u \in L^p(\mathbb{R}^n)$, then u is continuous and*

$$|u(x+h) - u(x)| \leq C_{n,p} \|\nabla u\|_p |h|^{1-\frac{n}{p}}, \qquad \textit{for every} \quad x, h \in \mathbb{R}^n, \tag{2.9}$$

for some positive constant $C_{n,p}$ depending only on p and n.

In particular $W^{1,p}(\mathbb{R}^n)$ is continuously embedded in the space of Hölder continuous functions $C^\beta(\mathbb{R}^n)$, with $\beta = 1 - \frac{n}{p}$.

Proof Consider a function $u \in C_0^\infty(\mathbb{R}^n)$, let $x, h \in \mathbb{R}^n$, and $v(x)$ be the function defined in (2.7). We next estimate I_A, I_B and I_C by using the Hölder inequality.

From the first inequality in (2.8) and (2.4), with $\alpha = n$ and $q = \dfrac{p}{p-1}$, we obtain

$$\begin{aligned} |I_A(x)| &\leq M|h| \|\nabla u\|_{L^p(\mathbb{R}^n)} \left\| \frac{1}{|z|^n} \right\|_{L^q(\{|z| \geq 2|h|\})} \\ &= M|h| \|\nabla u\|_{L^p(\mathbb{R}^n)} c_{n,q} (2|h|)^{\frac{n(p-1)}{p} - n} = M_A \|\nabla u\|_{L^p(\mathbb{R}^n)} |h|^{1-\frac{n}{p}}, \end{aligned}$$

for some positive constant M_A only depending on M, p and n. Moreover, from the second inequality in (2.8) and (2.3), with $\alpha = n-1$ and $q = \dfrac{p}{p-1}$, we find

$$\begin{aligned} |I_B(x)| &\leq c_n \|\nabla u\|_{L^p(\mathbb{R}^n)} \left\| \frac{1}{|z|^{n-1}} \right\|_{L^q(\{|z| \leq 2|h|\})} \\ &= c_n \|\nabla u\|_{L^p(\mathbb{R}^n)} c_{n-1,q} (2|h|)^{\frac{n(p-1)}{p} - n + 1} = M_B \|\nabla u\|_{L^p(\mathbb{R}^n)} |h|^{1-\frac{n}{p}}, \end{aligned}$$

where M_B is a positive constant only depending on p and n. Finally the last term in (2.7) can be estimated similarly to the second one, observing that $|x - y| \leq |x + h - y| + |h| < 3|h|$, thus getting

$$|I_C(x)| \leq c_n \|\nabla u\|_{L^p(\mathbb{R}^n)} \left\| \frac{1}{|z|^{n-1}} \right\|_{L^q(\{|z| \leq 3|h|\})} = M_C \|\nabla u\|_{L^p(\mathbb{R}^n)} |h|^{1-\frac{n}{p}},$$

for some positive constant M_C. We note that the L^q norms of the functions $z \mapsto |z|^{-n}$ and $z \mapsto |z|^{-n+1}$ appearing in the above estimates are finite thanks to the assumption $p > n$. Then (2.9) is obtained with $C_{n,p} := M_A + M_B + M_C$. This proves our claim for $u \in C_0^\infty(\mathbb{R}^n)$. The general case follows by a density argument. □

We next prove the compactness of the Sobolev embedding (1.11) for $p < q < p^*$ starting again from (2.7). Here we use the Young inequality instead of the Hölder inequality.

Theorem 2.5 *Let $p, q \geq 1$ be such that $p < q < p^* := \frac{np}{n-p}$. Then there exists a positive constant $C_{p,q}$ depending on n, p, q such that*

$$\|u(h + \cdot) - u\|_{L^q(\mathbb{R}^n)} \leq C_{p,q} \, \|\nabla u\|_{L^p(\mathbb{R}^n)} \, |h|^{n\left(\frac{1}{q} - \frac{1}{p^*}\right)}.$$

for every $u \in C_0^\infty(\mathbb{R}^n)$ and for any $h \in \mathbb{R}^n$. In particular, as long as $q < p^*$, we have

$$\|u(h + \cdot) - u\|_{L^q(\mathbb{R}^n)} \to 0 \qquad as \qquad |h| \to 0.$$

Proof Consider a function $u \in C_0^\infty(\mathbb{R}^n)$, let $x, h \in \mathbb{R}^n$, and $v(x)$ be the function defined in (2.7). We next estimate the L^q norm of I_A, I_B and I_C by using the Young inequality. To this aim we introduce the exponent r defined by the identity

$$1 + \frac{1}{q} = \frac{1}{r} + \frac{1}{p}, \tag{2.10}$$

and we note that

$$1 < r < \frac{n}{n-1} \quad \Leftrightarrow \quad p < q < p^*. \tag{2.11}$$

From the first inequality in (2.8) and (2.4), with $\alpha = n$, we obtain

$$\begin{aligned}\|I_A\|_{L^q(\mathbb{R}^n)} &\leq M|h| \|\nabla u\|_{L^p(\mathbb{R}^n)} \left\| \frac{1}{|z|^n} \right\|_{L^r(\{|z| \geq 2|h|\})} \\ &= M|h| \|\nabla u\|_{L^p(\mathbb{R}^n)} c_{n,r} (2|h|)^{\frac{n}{r} - n} = C_A(n, r) |h|^{n\left(\frac{1}{q} - \frac{1}{p^*}\right)},\end{aligned}$$

From the second inequality in (2.8) and (2.3), with $\alpha = n-1$, we obtain

$$\|I_B\|_{L^q(\mathbb{R}^n)} \leq c_n \|\nabla u\|_{L^p(\mathbb{R}^n)} \left\| \frac{1}{|z|^{n-1}} \right\|_{L^r(\{|z|\leq 2|h|\})}$$

$$= c_n \|\nabla u\|_{L^p(\mathbb{R}^n)} c_{n-1,r} (2|h|)^{\frac{n}{r}-n+1} = C_B(n,r) \|\nabla u\|_{L^p(\mathbb{R}^n)} |h|^{n\left(\frac{1}{q}-\frac{1}{p^*}\right)},$$

where $C_B(n,r)$ is a constant depending on n and r (and thus on n, p, q). The same argument applies to I_C, so that, provided that we consider the norm $\left\| \frac{1}{|z|^{n-1}} \right\|_{L^r(\{|z|\leq 3|h|\})}$ instead of $\left\| \frac{1}{|z|^{n-1}} \right\|_{L^r(\{|z|\leq 2|h|\})}$, as in the proof of the Morrey's theorem. We then find

$$\|I_C\|_{L^q(\mathbb{R}^n)} \leq C_C(n,r) \|\nabla u\|_{L^p(\mathbb{R}^n)} |h|^{n\left(\frac{1}{q}-\frac{1}{p^*}\right)},$$

We note that the L^r norms of the functions $z \mapsto |z|^{-n}$ and $z \mapsto |z|^{-n+1}$ appearing in the above estimates are finite if, and only if, the condition (2.11) is satisfied. The thesis is obtained with $C_{p,q} := C_A(n,r) + C_B(n,r) + C_C(n,r)$ for $u \in C_0^\infty(\mathbb{R}^n)$. The general case follows by a density argument. □

2.3 A More General Compactness Result

We note that the Theorem 2.5 only applies to a kernel that is homogeneous of degree $-n+1$. Actually, the method used in its proof also applies to any general homogeneous kernel K_α, with $0 < \alpha < n$, as those considered in Theorem 2.1. In the following statement we denote by u the convolution $K_\alpha * f$, that is

$$u(x) = \int_{\mathbb{R}^n} K_\alpha(x-y) f(y)\, dy.$$

Theorem 2.6 *Let K_α be a $C^1(\mathbb{R}^n \setminus \{0\})$ homogeneous function of degree $-\alpha$, with $1 < \alpha < n$, and let $p, q \geq 1$ be such that $q > p$ and*

$$1 - \frac{\alpha+1}{n} < \frac{1}{p} - \frac{1}{q} < 1 - \frac{\alpha}{n}. \tag{2.12}$$

Then there exists a positive constant $\widetilde{C}_{p,q}$, depending on n, p, q, such that

$$\|u(h+\cdot) - u\|_{L^q(\mathbb{R}^n)} \leq \widetilde{C}_{p,q}\, \|f\|_{L^p(\mathbb{R}^n)}\, |h|^{\frac{n}{r}-\alpha},$$

for every $f \in L^p(\mathbb{R}^n)$ and $h \in \mathbb{R}^n$. Here r is the constant introduced in (2.10), that is $1 + \frac{1}{q} = \frac{1}{r} + \frac{1}{p}$. Moreover the exponent $\frac{n}{r} - \alpha$ is strictly positive.

Proof We choose $x, h \in \mathbb{R}^n$ and let v be defined as in (2.6): $v(x) = u(x+h) - u(x)$, and we consider three integrals $v(x) = I_A(x) + I_B(x) + I_C(x)$ as in (2.7). We proceed as we did in the proof of Theorem 2.4. We find

$$
\begin{aligned}
|I_A(x)| &\leq M_\alpha |h| \int_{\{y \in \mathbb{R}^n : |x+h-y| \geq 2|h|\}} \frac{1}{|x+h-y|^{\alpha+1}} |f(y)| \, dy, \\
|I_B(x)| &\leq c_\alpha \int_{\{y \in \mathbb{R}^n : |x+h-y| < 2|h|\}} \frac{1}{|x+h-y|^{\alpha}} |f(y)| \, dy, \\
|I_C(x)| &\leq c_\alpha \int_{\{y \in \mathbb{R}^n : |x+h-y| < 2|h|\}} \frac{1}{|x-y|^{\alpha}} |f(y)| \, dy,
\end{aligned}
$$

being $M_\alpha := \max_{\frac{1}{2} \leq |z| \leq \frac{3}{2}} |\nabla K_\alpha(z)|$. In order to use the Young inequality, we recall that

$$
\left\| \frac{1}{|z|^{\alpha+1}} \right\|_{L^r(\{y \in \mathbb{R}^n : |z| \geq 2|h|\})} = c_{\alpha+1,r} (2|h|)^{\frac{n}{r} - (\alpha+1)},
$$

where r is the exponent introduced in (2.10) and we note that the above integral converges if, and only if, $(\alpha + 1)r > n$. Moreover,

$$
\left\| \frac{1}{|z|^{\alpha}} \right\|_{L^r(\{z \in \mathbb{R}^n : |z| \leq 2|h|\})} = c_{\alpha,r} (2|h|)^{\frac{n}{r} - \alpha},
$$

and the above integral converges if, and only if, $\alpha r < n$. Summarizing, the two above integrals are finite if, and only if,

$$
\frac{n}{\alpha + 1} < r < \frac{n}{\alpha}, \tag{2.13}
$$

which is equivalent to (2.12). We also note that the exponent $\frac{n}{r} - \alpha$ is strictly positive.

We next proceed as in the proof of Theorem 2.4. By the Young inequality we deduce

$$
\begin{aligned}
\|I_A\|_{L^q(\mathbb{R}^n)} &\leq \widetilde{C}_A(n, \alpha, r) |h|^{\left(\frac{n}{r} - \alpha\right)} \|f\|_{L^p(\mathbb{R}^n)}, \\
\|I_B\|_{L^q(\mathbb{R}^n)} &\leq \widetilde{C}_B(n, \alpha, r) |h|^{\left(\frac{n}{r} - \alpha\right)} \|f\|_{L^p(\mathbb{R}^n)}, \\
\|I_C\|_{L^q(\mathbb{R}^n)} &\leq \widetilde{C}_C(n, \alpha, r) |h|^{\left(\frac{n}{r} - \alpha\right)} \|f\|_{L^p(\mathbb{R}^n)},
\end{aligned}
$$

where $\widetilde{C}_A(n, \alpha, r)$, $\widetilde{C}_B(n, \alpha, r)$, and $\widetilde{C}_C(n, \alpha, r)$ are positive constants only depending on n, p, q, on c_α and on M_α. The thesis is obtained with $\widetilde{C}_{p,q} := \widetilde{C}_A(n, \alpha, r) + \widetilde{C}_B(n, \alpha, r) + \widetilde{C}_C(n, \alpha, r)$. □

Remark 2.7 The statement of Theorem 2.6 is more involved with respect to that of Theorem 2.5 as we don't have a natural counterpart of the Sobolev exponent p^* for any $\alpha \in]0, n[$. We list here the explicit conditions on q for the validity of (2.12). We discuss specifically the case $1 < p < n$ as we are interested in the compactness of the Sobolev embedding.

(i) If $\boxed{0 < \alpha < n-2}$ then

(i.1) If $1 < p < \dfrac{n}{n-\alpha}$ then

$$\frac{np}{n-p(n-1-\alpha)} < q < \frac{np}{n-p(n-\alpha)}$$

(i.2) If $\dfrac{n}{n-\alpha} < p < \dfrac{n}{n-1-\alpha}$ then

$$q > \max\left\{1, \frac{np}{n-p(n-1-\alpha)}\right\} = \frac{np}{n-p(n-\alpha)}$$

(i.3) If $\dfrac{n}{n-1-\alpha} < p < n$ then no values of q are available

(ii) If $\boxed{\alpha = n-2}$ then

(ii.1) If $1 < p < \dfrac{n}{2}$ then

$$p^* = \frac{np}{n-p} < q < \frac{np}{n-2p}$$

(ii.2) If $\dfrac{n}{2} < p < n$ then

$$q > \max\left\{1, \frac{np}{n-p}\right\} = \frac{np}{n-p} = p^*$$

(iii) If $\boxed{n-2 < \alpha < n-1}$ then

(iii.1) If $1 < p < \dfrac{n}{n-\alpha}$ then

$$\frac{np}{n-p(n-1-\alpha)} < q < \frac{np}{n-p(n-\alpha)}$$

(iii.2) If $\dfrac{n}{n-\alpha} < p < n$ then

$$q > \max\left\{1, \frac{np}{n-p(n-\alpha)}\right\} = \frac{np}{n-p(n-1-\alpha)}$$

(iv) If $\boxed{\alpha = n-1}$ then for any $1 < p < n$

$$p < q < p^* = \frac{np}{n-p}$$

(v) If $\boxed{n-1 < \alpha < n}$ then for any $1 < p < n$

$$\frac{np}{n-p(n-1-\alpha)} < q < \frac{np}{n-p(n-\alpha)}.$$

3 Continuous and Compact Embeddings for Degenerate Kolmogorov Equations

The aim of this section is to prove compactness estimates for weak solutions to a family of degenerate Kolmogorov operators that includes the one in (1.19) as the simplest prototype. Specifically, we consider operators in this form

$$\mathscr{L}u(x,t) := \sum_{i,j=1}^{m_0} \partial_{x_i}(a_{i,j}(x,t)\partial_{x_j}u(x,t)) + \sum_{i,j=1}^{N} b_{i,j}x_i\partial_{x_j}u(x,t) - \partial_t u(x,t), \tag{3.1}$$

with $(x,t) = (x_1, \dots, x_N, t) \in \mathbb{R}^{N+1}$. Here $m_0 \in \mathbb{N}$ is such that $1 \leq m_0 \leq N$. In the sequel we will also use the following notation $z := (x,t)$, and we always assume the following hypotheses:

(H.1) The matrix $A = (a_{i,j}(z))_{i,j=1,\dots,m_0}$ is symmetric, with measurable coefficients and there exists a positive constant μ such that

$$\mu^{-1}|\xi|^2 \leq \sum_{i,j=1}^{m_0} a_{i,j}(z)\xi_i\xi_j \leq \mu|\xi|^2$$

for all $z \in \mathbb{R}^{N+1}$ and $\xi \in \mathbb{R}^{m_0}$.

(H.2) The matrix B has constant coefficients. Moreover there exists a basis of $\mathbb{R}^N$ such that the matrix B can be written in a canonical form:

$$B = \begin{pmatrix} 0 & B_1 & 0 & \dots & 0 \\ 0 & 0 & B_2 & \dots & 0 \\ \vdots & \vdots & \vdots & \ddots & \vdots \\ 0 & 0 & 0 & \dots & B_r \\ 0 & 0 & 0 & \dots & 0 \end{pmatrix}$$

where B_k is a matrix $m_{k-1} \times m_k$ with rank m_k, $k = 1, 2, \dots, r$ with

$$m_0 \geq m_1 \geq \dots \geq m_r \geq 1, \quad \text{e} \quad \sum_{k=0}^{r} m_k = N.$$

We prove in detail Theorem 1.1 along the same techniques outlined in the Introduction for the operator $\mathscr{L}$ in (1.19). In particular, we will rely on a representation formula analogous to (1.2) in terms of the fundamental solution to the operator

$$\mathscr{L}_0 u := \sum_{i=1}^{m_0} \partial_{x_i}^2 u + \sum_{i,j=1}^{N} b_{i,j} x_i \partial_{x_j} u - \partial_t u. \tag{3.2}$$

Remark 1 It is know that Assumption (H.2) is equivalent to the assumption of hypoellipticity of $\mathscr{L}_0$ (see [7] and its bibliography). This means that any function u which is a distributional solution to $\mathscr{L}_0 u = f$ in some open subset Ω of $\mathbb{R}^{N+1}$ is a C^∞ function whenever f is C^∞.

In order to simplify our statements, in the sequel we adopt the following compact notation. If I_{m_0} is the identity matrix $m_0 \times m_0$, we set

$$\Delta_{m_0} = \sum_{i=1}^{m_0} \partial_{x_i}^2, \qquad Y = \sum_{i,j=1}^{N} b_{i,j} x_i \partial_{x_j} - \partial_t, \qquad A_0 = \begin{pmatrix} I_{m_0} & 0 \\ 0 & 0 \end{pmatrix}.$$

In particular we will write

$$\mathscr{L}_0 = \Delta_{m_0} + Y = \operatorname{div}(A_0 \nabla) + Y.$$

We next introduce the non-Euclidean geometric setting suitable for the study of $\mathscr{L}$, the fundamental solution of $\mathscr{L}_0$ and the definition of the convolution with homogeneous kernels.

3.1 Dilation and Translation Groups Associated to $\mathscr{L}$

We recall here some invariance properties of the operator $\mathscr{L}_0$. We refer to [7] where the definition of translation group and dilation group for Kolmogorov operators have been given for the first time. Let

$$(x,t)\circ(\xi,\tau) = (\xi + E(\tau)x, t+\tau), \quad E(t) = \exp(-tB^T), \; (x,t), (\xi,\tau) \in \mathbb{R}^{N+1} \tag{3.3}$$

$$D(\lambda) = \mathrm{diag}(\lambda I_{m_0}, \lambda^3 I_{m_1}, \ldots, \lambda^{2r+1} I_{m_r}, \lambda^2), \quad \lambda > 0 \tag{3.4}$$

where I_{m_j} denotes the identity matrix $m_j \times m_j$. It is known that $(\mathbb{R}^{N+1}, \circ)$ is a *non commutative* group, and $\mathscr{L}_0$ is invariant with respect to the *left translations* of $(\mathbb{R}^{N+1}, \circ)$, in the following sense: if we choose any $\zeta \in \mathbb{R}^{N+1}$ and we set $v(z) := u(\zeta \circ z)$ and $g(z) := f(\zeta \circ z)$, then the have

$$\mathscr{L}_0 u(z) = f(z) \quad \Leftrightarrow \quad \mathscr{L}_0 v(z) = g(z).$$

Moreover, $\mathscr{L}_0$ is invariant with respect to $(D(\lambda))_{\lambda>0}$, with the following meaning:

$$\mathscr{L}_0 u(z) = f(z) \quad \Leftrightarrow \quad \mathscr{L}_0 w(z) = \lambda^2 h(z).$$

Now $w(z) := u(D(\lambda)z)$, $h(z) := f(D(\lambda)z)$, and λ is any positive constant. The zero of the group is $(0,0)$ and the inverse of (ξ,τ) is $(\xi,\tau)^{-1} = (-E(\tau)\xi, -\tau)$. Moreover the following *distributive property* holds:

$$D(\lambda)\,(z \circ \zeta) = (D(\lambda)z) \circ (D(\lambda)\zeta)$$

We summarize the above properties by saying that $\mathscr{L}_0$ is invariant with respect to the *homogeneous group* $\left(\mathbb{R}^{N+1}, \circ, (D(\lambda))_{\lambda>0}\right)$.

In the sequel we will use the following notation

$$D(\lambda) = \mathrm{diag}(\lambda^{\alpha_1}, \lambda^{\alpha_2}, \ldots, \lambda^{\alpha_N}, \lambda^2),$$

where, in accordance with (3.4) $\alpha_1 = \alpha_2 = \cdots = \alpha_{m_0} = 1$, $\alpha_{m_0+1} = \cdots = \alpha_{m_0+m_1} = 3, \ldots, \alpha_{m_0+m_1+\cdots+m_{r-1}+1} = \cdots = \alpha_N = 2r+1$.

We define now a norm of $\mathbb{R}^{N+1}$ homogeneous of degree one with respect to the dilation introduced before.

Definition 3.1 For all $z \in \mathbb{R}^{N+1} \setminus \{0\}$, we define the norm $\|z\| = \rho$, as the unique positive solution to:

$$\frac{x_1^2}{\rho^{2\alpha_1}} + \frac{x_2^2}{\rho^{2\alpha_2}} + \ldots + \frac{x_N^2}{\rho^{2\alpha_N}} + \frac{t^2}{\rho^4} = 1$$

and $\|0\| = 0$.

This norm is homogeneous with respect to the dilation group $(D(\lambda))_{\lambda>0}$ as long as the following property holds:

$$\|D(\lambda)z\| = \lambda\|z\| \quad \forall z \in \mathbb{R}^{N+1} \setminus \{0\} \text{ and } \lambda > 0.$$

Moreover the following quasi-triangular inequality holds. There exists a constant $c_T \geq 1$ such that

$$\|z \circ \zeta\| \leq c_T(\|z\| + \|\zeta\|), \qquad \|z^{-1}\| \leq c_T\|z\|, \tag{3.5}$$

for every $z, \zeta \in \mathbb{R}^{N+1}$. We denote by $d(\zeta, z) := \|z^{-1} \circ \zeta\|$ the *quasi-distance* of z and ζ. We denote by $B_\varrho(z)$ the open ball of radius ϱ and center z with respect to the above quasi-distance

$$B_\varrho(z) := \left\{\zeta \in \mathbb{R}^{N+1} \mid \|z^{-1} \circ \zeta\| < \varrho\right\}. \tag{3.6}$$

Note that the topology induced on $\mathbb{R}^{N+1}$ from the norm introduced in Definition 3.1 is equivalent to the Euclidean one.

Remark 3.2 The Lebesgue measure is invariant with respect to the group $(\mathbb{R}^{N+1}, \circ)$. Moreover, as long as $\det D(\lambda) = \lambda^{Q+2}$, where

$$Q = m_0 + 3m_1 + \ldots + (2r+1)m_r$$

we have that the following identity holds:

$$\text{meas}(B_r(0)) = r^{Q+2}\text{meas}(B_1(0)),$$

where $\text{meas}(B)$ indicates the Lebesgue measure of the set B. For this reason, we will refer to $(Q+2)$ as the *homogeneous dimension* of $\mathbb{R}^{N+1}$ with respect to the dilation (3.4).

We also note that, in view of the structure of the matrix B and of the definition of $E(\tau)$, we have that $\det E(\tau) = 1$ for every τ. In particular, the Jacobian determinant of the left translation $(x,t) \mapsto (\xi,\tau) \circ (x,t)$ agrees with 1 for every $(\xi,\tau) \in \mathbb{R}^{N+1}$. The same is true for the Jacobian determinant of $(\xi,\tau) \mapsto (\xi,\tau)^{-1}$. As a consequence we have that

$$\begin{aligned}
&\int_A f(\zeta \circ z)dz = \int_{\zeta\circ A} f(w)dw, && \zeta \circ A := \left\{w = \zeta \circ z \mid z \in A\right\},\\
&\int_A f(z^{-1})dz = \int_{A^{-1}} f(w)dw, && A^{-1} := \left\{w = z^{-1} \mid z \in A\right\},\\
&\int_A f(z^{-1} \circ \zeta)dz = \int_{A^{-1}\circ\zeta} f(w)dw, && A^{-1} \circ \zeta := \left\{w = z^{-1} \circ \zeta \mid z \in A\right\}.
\end{aligned} \tag{3.7}$$

The following results is the analogous of (2.3) and (2.4) in the setting of the homogeneous Lie group $\left(\mathbb{R}^{N+1}, \circ, (D(\lambda))_{\lambda>0}\right)$.

Lemma 3.3 *Let K_α denote any continuous function which is homogeneous of degree $-\alpha$ with respect to $(D(\lambda))_{\lambda>0}$, for some α such that $0 < \alpha < Q+2$. For every $R > 0$ we have that:*

- $K_\alpha \in L^q\left(\{z \in \mathbb{R}^{N+1} \mid \|z\| \le R\}\right)$ *if, and only if,* $q > \dfrac{Q+2}{\alpha}$. *Moreover, there exists a positive constant $\widetilde{c}_{\alpha,q}$, only depending on K_α, $(D(\lambda))_{\lambda>0}$ and q, such that*

$$\|K_\alpha\|_{L^q(\{z\in\mathbb{R}^{N+1} \mid \|z\|\le R\})} \le \widetilde{c}_{\alpha,q} R^{\frac{Q+2}{q}-\alpha}, \tag{3.8}$$

- $K_\alpha \in L^q\left(\{z \in \mathbb{R}^{N+1} \mid \|z\| \ge R\}\right)$ *if, and only if,* $q < \dfrac{Q+2}{\alpha}$. *Moreover, there exists a positive constant $\widetilde{c}_{\alpha,q}$, only depending on K_α, $(D(\lambda))_{\lambda>0}$ and q, such that*

$$\|K_\alpha\|_{L^q(\{z\in\mathbb{R}^{N+1} \mid \|z\|\ge R\})} \le \widetilde{c}_{\alpha,q} R^{\frac{Q+2}{q}-\alpha}. \tag{3.9}$$

Proof We compute the integrals by using the "polar coordinates"

$$\begin{cases} x_1 = \rho^{\alpha_1} \cos\psi_1 \dots \cos\psi_{N-1} \cos\psi_N \\ x_2 = \rho^{\alpha_2} \cos\psi_1 \dots \cos\psi_{N-1} \sin\psi_N \\ \vdots \\ x_N = \rho^{\alpha_N} \cos\psi_1 \sin\psi_2 \\ t = \rho^2 \sin\psi_1. \end{cases}$$

Note that, in accordance with the Definition 3.1, the Jacobian determinant of the above change of coordinate is homogeneous of degree $Q+1$ with respect to the variable ρ, that is $J(\rho,\psi) = \rho^{Q+1} J(1,\psi)$. The claim then follows by proceedings as in the Euclidean case. □

3.2 Preliminary Results on Convolutions in Homogeneous Lie Groups

We recall some facts concerning the convolution of functions in homogeneous Lie groups. We refer to the work of Folland [6], and to its bibliography, for a comprehensive treatment of this subject. The first result is a Young inequality in the non-Euclidean setting

Theorem 3.4 *Let $p, q, r \in [1, +\infty]$ be such that:*

$$1 + \frac{1}{q} = \frac{1}{p} + \frac{1}{r}. \tag{3.10}$$

If $f \in L^p(\mathbb{R}^{N+1})$ and $g \in L^r(\mathbb{R}^{N+1})$, then the function $f * g$ defined as:

$$f * g(z) := \int_{\mathbb{R}^{N+1}} f(\zeta^{-1} \circ z) g(\zeta)\, \mathrm{d}\zeta$$

belongs to $L^q(\mathbb{R}^{N+1})$ and it holds:

$$\|f * g\|_{L^q(\mathbb{R}^{N+1})} \le \|f\|_{L^p(\mathbb{R}^{N+1})} \|g\|_{L^r(\mathbb{R}^{N+1})}.$$

The following two theorems are the counterpart of Theorem 2.1 and Lemma 2.2 in Sect. 2, respectively.

Theorem 3.5 *(Proposition (1.11) in [6]) Let K_α be a continuous function, homogeneous of degree $-\alpha$ with $0 < \alpha < Q+2$, with respect to the dilation (3.4). Then, for every $p \in]1, +\infty[$, the convolution u of K_α with a function $f \in L^p(\mathbb{R}^{N+1})$*

$$u(z) = \int_{\mathbb{R}^{N+1}} K_\alpha(\zeta^{-1} \circ z) f(\zeta)\, \mathrm{d}\zeta, \tag{3.11}$$

is defined for almost every $z \in \mathbb{R}^{N+1}$ and is a measurable function. Moreover there exists a constant $\widehat{C}_p = \widehat{C}_p(p, Q)$ such that

$$\|u\|_{L^q(\mathbb{R}^{N+1})} \le \widehat{C}_p \max_{\|z\|=1} |K_\alpha(z)| \|f\|_{L^p(\mathbb{R}^{N+1})},$$

for every $f \in L^p(\mathbb{R}^{N+1})$, where q is defined by

$$1 + \frac{1}{q} = \frac{1}{p} + \frac{\alpha}{Q+2}.$$

For the proof of the next result we refer to Proposition (1.11) in [6] or Lemma 5.1 in [8].

Theorem 3.6 *Let $K_\alpha \in C^1(\mathbb{R}^{N+1} \setminus \{0\})$ be a homogeneous function of degree $-\alpha$ with respect to the group $(D(\lambda))_{\lambda>0}$. Then there exist two constants $\kappa > 1$ and*

$M_\alpha > 0$ such that:

$$|K_\alpha(\zeta) - K_\alpha(z)| \le M_\alpha \frac{\|z^{-1} \circ \zeta\|}{\|z\|^{\alpha+1}}$$

for all z, ζ such that $\|z\| \ge \kappa \|z^{-1} \circ \zeta\|$.

3.3 Compactness Estimates for Convolutions with Homogeneous Kernels

Theorem 3.7 *Let K_α be a $C^1(\mathbb{R}^{N+1} \setminus \{0\})$ homogeneous function of degree $-\alpha$ with $0 < \alpha < Q + 2$, with respect to the dilation (3.4). Then for every $p, q \ge 1$ such that $q > p$ and*

$$1 - \frac{\alpha + 1}{Q + 2} < \frac{1}{p} - \frac{1}{q} < 1 - \frac{\alpha}{Q + 2}, \tag{3.12}$$

there exists a positive constant $\widetilde{C}_{p,q}$ depending on α, p, q and on the dilation group $(D(\lambda))_{\lambda>0}$ such that

$$\|u(\cdot \circ h) - u\|_{L^q(\mathbb{R}^{N+1})} \le \widetilde{C}_{p,q} \|h\|^{\frac{Q+2}{r} - \alpha} \|f\|_{L^p(\mathbb{R}^{N+1})},$$

for every $f \in L^p(\mathbb{R}^{N+1})$, and $h \in \mathbb{R}^{N+1}$. Here r is the constant defined by (3.10), that is

$$1 + \frac{1}{q} = \frac{1}{r} + \frac{1}{p},$$

and the exponent $\frac{Q+2}{r} - \alpha$ is strictly positive, because of (3.12).

Proof We proceed as in the proof of Theorem 2.6. We choose $z, h \in \mathbb{R}^{N+1}$ and we let

$$v(z) := u(z \circ h) - u(z). \tag{3.13}$$

By the formula (3.11) we have $v(z) = \widetilde{I}_A(z) + \widetilde{I}_B(z) + \widetilde{I}_C(z)$, where

$$\begin{aligned}
\widetilde{I}_A(z) &= \int_{\{\zeta \in \mathbb{R}^{N+1}: \|\zeta^{-1} \circ z \circ h\| \ge \kappa \|h\|\}} \left(K_\alpha(\zeta^{-1} \circ z \circ h) - K_\alpha(\zeta^{-1} \circ z) \right) f(\zeta)\, d\zeta, \\
\widetilde{I}_B(z) &= \int_{\{\zeta \in \mathbb{R}^{N+1}: \|\zeta^{-1} \circ z \circ h\| < \kappa \|h\|\}} K_\alpha(\zeta^{-1} \circ z \circ h) f(\zeta)\, d\zeta, \\
\widetilde{I}_C(z) &= \int_{\{\zeta \in \mathbb{R}^{N+1}: \|\zeta^{-1} \circ z \circ h\| < \kappa \|h\|\}} -K_\alpha(\zeta^{-1} \circ z) f(\zeta)\, d\zeta.
\end{aligned} \tag{3.14}$$

Then, as in the proof of Theorem 2.6, we find

$$|\widetilde{I}_A(z)| \le c_T M_\alpha \|h\| \int_{\{\zeta\in\mathbb{R}^{N+1}:\|\zeta^{-1}\circ z\circ h\|\ge\kappa\|h\|\}} \frac{1}{\|\zeta^{-1}\circ z\circ h\|^{\alpha+1}} |f(\zeta)|\, d\zeta,$$

$$|\widetilde{I}_B(z)| \le c_\alpha \int_{\{\zeta\in\mathbb{R}^{N+1}:\|\zeta^{-1}\circ z\circ h\|<\kappa\|h\|\}} \frac{1}{\|\zeta^{-1}\circ z\circ h\|^{\alpha}} |f(\zeta)|\, d\zeta,$$

$$|\widetilde{I}_C(z)| \le c_\alpha \int_{\{\zeta\in\mathbb{R}^{N+1}:\|\zeta^{-1}\circ z\circ h\|<\kappa\|h\|\}} \frac{1}{\|\zeta^{-1}\circ z\|^{\alpha}} |f(\zeta)|\, d\zeta.$$

The first estimate follows from Theorem 3.6 and the constant c_T is the one appearing in (3.5), while $c_\alpha := \max_{\|w\|=1} K_\alpha(w)$.

We next compute the L^r norm of the homogeneous functions appearing above. In view of Remark 3.2 and Lemma 3.3, we have that

$$\left\| \frac{1}{\|\zeta^{-1}\circ z\circ h\|^{\alpha+1}} \right\|_{L^r(\{\zeta\in\mathbb{R}^{N+1}:\|\zeta^{-1}\circ z\circ h\|\ge\kappa\|h\|\})} = \widetilde{C}_A(r,Q)\|h\|^{\frac{Q+2}{r}-(\alpha+1)},$$

where r is the exponent introduced in (2.10) and $\widetilde{C}_A(r,Q)$ is a constant depending on Q and r (hence on Q, p, q), and on the dilation group $(D(\lambda))_{\lambda>0}$. Using again Remark 3.2 and Lemma 3.3, and we also find

$$\left\| \frac{1}{\|\zeta^{-1}\circ z\circ h\|^{\alpha}} \right\|_{L^r(\{\zeta\in\mathbb{R}^{N+1}:\|\zeta^{-1}\circ z\circ h\|\le\kappa\|h\|\})} = \widetilde{C}_B(r,Q)\|h\|^{\frac{Q+2}{r}-\alpha},$$

The same argument applies to $\widetilde{I}_C$, by using the quasi-triangular inequality (3.5), so that

$$\left\| \frac{1}{\|\zeta^{-1}\circ z\circ h\|^{\alpha}} \right\|_{L^r(\{\zeta\in\mathbb{R}^{N+1}:\|\zeta^{-1}\circ z\circ h\|\le\kappa\|h\|\})} = \widetilde{C}_C(r,Q)\|h\|^{\frac{Q+2}{r}-\alpha},$$

Note that the three above integrals converge if, and only if,

$$\frac{Q+2}{\alpha+1} < r < \frac{Q+2}{\alpha}. \tag{3.15}$$

We note that (3.12) is equivalent to (3.15) and that the second inequality in (3.15) says that the exponent $\frac{Q+2}{r} - \alpha$ appearing in the statement of this Theorem is strictly positive. By the Young inequality (Theorem 3.4) we conclude that

$$\|\widetilde{I}_A\|_{L^q(\mathbb{R}^{N+1})} \le c_T M_\alpha \widetilde{C}_A(r,Q) \|h\|^{\frac{Q+2}{r}-\alpha} \|f\|_{L^p(\mathbb{R}^{N+1})},$$

$$\|\widetilde{I}_B\|_{L^q(\mathbb{R}^{N+1})} \le c_\alpha \widetilde{C}_B(r,Q) \|h\|^{\frac{Q+2}{r}-\alpha} \|f\|_{L^p(\mathbb{R}^{N+1})},$$

$$\|\widetilde{I}_B\|_{L^q(\mathbb{R}^{N+1})} \le c_\alpha \widetilde{C}_C(r,Q) \|h\|^{\frac{Q+2}{r}-\alpha} \|f\|_{L^p(\mathbb{R}^{N+1})}.$$

The thesis is obtained with $\widetilde{C}_{p,q} := c_T M_\alpha \widetilde{C}_A(r,Q) + c_\alpha \widetilde{C}_B(r,Q) + c_\alpha \widetilde{C}_C(r,Q)$. □

Remark 3.8 Similarly as we did in Remark 2.7, we can state the conditions on q for the validity of (3.12). They can be obtained by substituting the *dimension* n with the *homogeneous dimension* $Q+2$. We explicitly write here the condition for $\alpha = Q$ and $\alpha = Q+1$, which occur in the representation formulas for the solutions to $\mathscr{L}u = f$. Moreover, when $\alpha = Q+1$, we only consider the case $1 < p < Q+2$, as we apply Theorem 3.7 to prove the compactness of the embedding of Theorem 3.5, which holds only for $p < Q+2$. For the same reason, when $\alpha = Q$, we only consider the case $1 < p < \frac{Q+2}{2}$.

(i) If $\alpha = Q$ and $1 < p < \frac{Q+2}{2}$, we have that $p^* < q < p^{**}$,
(ii) If $\alpha = Q+1$ and $1 < p < Q+2$ we have that $p < q < p^*$.

3.4 Proof of Our Main Result

Proof of Theorem 1.1 The proof of the Sobolev inequality is a direct consequence of Theorem 3.5, with $\alpha = Q+1$ when considering $v_1, \dots, v_{m_0}$, and $\alpha = Q$ as we consider u.

The compactness of the embedding is a direct consequence of Theorem 3.7. As noticed in Remark 3.8, it applies to the derivatives $\partial_{\xi_j}\Gamma$, that are homogeneous functions of degree $-(Q+1)$, only when $p < q < p^*$. Moreover, a direct computation based on (3.10) shows that

$$\frac{Q+2}{r} - (Q+1) = (Q+2)\left(\frac{1}{q} - \frac{1}{p^*}\right). \tag{3.16}$$

Analogously, as Γ is homogeneous of degree $-Q$, we need to consider $p^* < q < p^{**}$. In this case, by using again (3.10) we find

$$\frac{Q+2}{r} - Q = (Q+2)\left(\frac{1}{q} - \frac{1}{p^{**}}\right). \tag{3.17}$$

The proof of the Morrey embedding is obtained by the same argument used in the proof of Theorem 2.4. We consider the function $v(z) = u(z \circ h) - u(z)$ introduced in (3.13) and, as in the proof of Theorem 3.7, we write $v(z) = \widetilde{I}_A(z) + \widetilde{I}_B(z) + \widetilde{I}_C(z)$, where the functions $\widetilde{I}_A, \widetilde{I}_B, \widetilde{I}_C$ are defined in (3.14). The conclusion of the proof is obtained by using the Young inequality stated in Theorem 3.4. □

4 Representation Formulas

4.1 Fundamental Solution to $\mathscr{L}_0$ and Representation Formula

In this Section we focus on the representation formulas for the equation $\mathscr{L}u = 0$, for $\mathscr{L}$ satisfying the assumptions (H.1) and (H.2), then we prove Theorem 1.2.

We first recall the definition of *weak solution* to $\mathscr{L}u = 0$, then we recall that, under these assumptions, the fundamental solution to $\mathscr{L}_0$ has been derived by Hörmander in [6]. We say that u is a weak solution to $\mathscr{L}u = 0$ in an open set $\Omega \subset \mathbb{R}^{N+1}$ if $u, \partial_{x_1}u, \ldots, \partial_{x_{m_0}}u, Yu \in L^2_{\text{loc}}(\Omega)$ and

$$\int_\Omega -\langle A(z)\nabla u(z), \nabla\psi(z)\rangle + \psi(z)Yu(z)\,\mathrm{d}z = 0,$$

for all $\psi \in C_0^\infty(\Omega)$.

With the notations introduced in Sect. 3, we let

$$C(t) = \int_0^t E(s)A_0E^T(s)\,\mathrm{d}s$$

The assumptions (H.1) and (H.2) guarantee that the matrix $C(t)$ is strictly positive for all $t > 0$. In this case its inverse $C^{-1}(t)$ is well defined and the fundamental solution to $\mathscr{L}_0$ with singularity at the origin of $\mathbb{R}^{N+1}$, is given by

$$\Gamma((x,t),(0,0)) = \begin{cases} \dfrac{(4\pi)^{-\frac{N}{2}}}{\sqrt{\det C(t)}} \exp\left(-\frac{1}{4}\left\langle C^{-1}(t)x, x\right\rangle\right), & \text{if } t > 0, \\ 0, & \text{if } t \le 0. \end{cases} \tag{4.1}$$

To simplify the notation, in the sequel we will write $\Gamma(x,t,\xi,\tau)$ instead of $\Gamma((x,t),(\xi,\tau))$, and $\Gamma(x,t)$ instead of $\Gamma(x,t,0,0)$. The fundamental solution $\Gamma(x,t,\xi,\tau)$ of $\mathscr{L}_0$ with pole at (ξ,τ) is the "left translation" of $\Gamma(\cdot,0,0)$ with respect to the group $(\mathbb{R}^{N+1},\circ)$:

$$\Gamma(x,t,\xi,\tau) = \Gamma((\xi,\tau)^{-1} \circ (x,t), 0, 0).$$

Let us explicitly remark that $\Gamma(\cdot, 0, 0)$ is homogeneous of degree $-Q$ with respect to the group $(D(\lambda))_{\lambda>0}$ and $\partial_{x_j}\Gamma(\cdot, 0, 0)$ is homogeneous of degree $-Q-1$, for $j = 1, \dots, m_0$. Moreover, also $\partial_{\xi_j}\Gamma(0, 0, \cdot)$ is homogeneous of degree $-Q-1$, for $j = 1, \dots, m_0$.

We next represent weak solutions to $\mathscr{L}u = 0$ as convolutions with the fundamental solution Γ to $\mathscr{L}_0$ and to its derivatives $\partial_{\xi_1}\Gamma, \dots, \partial_{\xi_{m_0}}\Gamma$.

Consider any open set $\Omega \subseteq \mathbb{R}^{N+1}$ let u be a function such that $u, \partial_{x_1}u, \dots, \partial_{x_{m_0}}u, Yu \in L^p_{\text{loc}}(\Omega)$, and $\eta \in C_0^\infty(\Omega)$ is any cut-off function, then, by an elementary density argument we find

$$(\eta u)(z) = -\int_{\mathbb{R}^{N+1}} [\Gamma(z, \cdot)\mathscr{L}_0(\eta u)](\zeta)\, \mathrm{d}\zeta =$$

$$\int_{\mathbb{R}^{N+1}} [\langle A_0 \nabla_\xi \Gamma(z, \cdot), \nabla(\eta u)\rangle](\zeta)\, \mathrm{d}\zeta - \int_{\mathbb{R}^{N+1}} [\Gamma(z, \cdot) Y(\eta u)](\zeta)\, \mathrm{d}\zeta.$$

If moreover u is a weak solutions to $\mathscr{L}u = \operatorname{div}(A_0F) + f$, then $\mathscr{L}_0 u = \operatorname{div}((A_0 - A)\nabla u + A_0 F) + f$, for some $f \in L^p_{\text{loc}}(\Omega)$ and some vector valued function $F = (F_1, \dots, F_{m_0}, 0, \dots, 0)$ with $F_1, \dots, F_{m_0} \in L^p_{\text{loc}}(\Omega)$, then we obtain the following representation formula introduced in Theorem 3.1 of [8], and used in the proof of Theorem 3.3 of [9].

Theorem 4.1 *If u is a weak solution to $\mathscr{L}u = \operatorname{div}(A_0F) + f$, in some open set $\Omega \subset \mathbb{R}^{N+1}$, with $f, F_1, \dots, F_{m_0} \in L^p_{\text{loc}}(\Omega)$, and η is the cut-off function defined above, then:*

$$(\eta u)(z) = \int_{\mathbb{R}^{N+1}} [\eta\langle \nabla_\xi \Gamma(z, \cdot), (A_0 - A)\nabla u + A_0 F\rangle](\zeta)\, \mathrm{d}\zeta - \int_{\mathbb{R}^{N+1}} [\Gamma(z, \cdot)(\langle A\nabla \eta, \nabla u\rangle + \eta f)]\, \mathrm{d}\zeta$$
$$+ \int_{\mathbb{R}^{N+1}} [\langle A_0 \nabla_\xi \Gamma(z, \cdot), \nabla \eta\rangle u](\zeta)\, \mathrm{d}\zeta - \int_{\mathbb{R}^{N+1}} \Gamma(z, \cdot)(Y\eta) u\, \mathrm{d}\zeta. \tag{4.2}$$

In the following statement $B_\varrho(z_0)$ denotes the ball defined in (3.6), and c_T is the constant in (3.5).

Proposition 4.2 *Let Ω be an open set of $\mathbb{R}^{N+1}$, and let u be a weak solution to $\mathscr{L}u = \operatorname{div}(A_0F) + f$ in Ω. Suppose that $u, f, \partial_{x_1}u, \dots, \partial_{x_{m_0}}u, F_1, \dots, F_{m_0} \in L^p_{\text{loc}}(\Omega)$. Then for every $z_0 \in \Omega$, and $\varrho, \sigma > 0$ such that the ball $B_\varrho(z_0)$ is contained in Ω, and $\sigma < \dfrac{\varrho}{2c_T}$, we have:*

- *(Sobolev embedding) if $1 < p < Q + 2$, then there exists a positive constant C_p such that*

$$\|u\|_{L^{p^*}(B_\sigma(z_0))} \le C_p\big(\|u\|_{L^p(B_\varrho(z_0))} + \|f\|_{L^p(B_\varrho(z_0))}$$
$$+ \|A_0 \nabla u\|_{L^p(B_\varrho(z_0))} + \|A_0 F\|_{L^p(B_\varrho(z_0))}\big),$$

- *(Compactness) if moreover $p < q < p^*$, then there exists a positive constant $\widetilde{C}_{p,q}$ such that*

$$\|u(\cdot \circ h) - u\|_{L^q(B_\sigma(z_0))} \leq \widetilde{C}_{p,q}\big(\|u\|_{L^p(B_\varrho(z_0))} + \|f\|_{L^p(B_\varrho(z_0))} + \|A_0 \nabla u\|_{L^p(B_\varrho(z_0))} + \|A_0 F\|_{L^p(B_\varrho(z_0))}\big)\, \|h\|^{(Q+2)\left(\frac{1}{q} - \frac{1}{p^*}\right)},$$

for every $h \in B_\sigma(z_0)$,

- *(Morrey embedding) if $p > Q+2$, then there exists a positive constant $\widetilde{C}_p$ such that*

$$|u(z) - u(\zeta)| \leq \widetilde{C}_p\big(\|u\|_{L^p(B_\varrho(z_0))} + \|f\|_{L^p(B_\varrho(z_0))} + \|A_0 \nabla u\|_{L^p(B_\varrho(z_0))} + \|A_0 F\|_{L^p(B_\varrho(z_0))}\big)\|\zeta^{-1} \circ z\|^{1-\frac{Q+2}{p}},$$

for every $z, \zeta \in B_\sigma(z_0)$.

Proof We apply Theorem 4.1 with a function η supported in the ball $B_\varrho(z_0)$ and such that $\psi(z) = 1$ for every $z \in B_{2c_T\sigma}(z_0)$. It is not difficult to check that a cut-off function with the above properties exists (see formula (3.3) in [8] for instance). Note that the integrals appearing in the Eq. (4.2) involving $\partial_{x_1} u, \ldots, \partial_{x_{m_0}} u, F_1, \ldots, F_{m_0}$ are convolutions of $\partial_{\xi_1}\Gamma, \ldots, \partial_{\xi_{m_0}}\Gamma$, that are homogeneous kernels of degree $-(Q+1)$, with functions belonging to $L^p(B_\varrho(z_0))$ multiplied by bounded functions compactly supported in $B_\varrho(z_0)$. For these terms of the representation formula (4.2) the thesis then follows from a direct application of Theorem 1.1. Indeed, by our choice of η, we have $u(z) = (\eta u)(z)$ for every $z \in B_\sigma(z_0)$. Moreover, if $z, h \in B_\sigma(z_0)$ we also have that $z \circ h \in B_{2c_T\sigma}(z_0)$, by (3.5), then also $u(z \circ h) = (\eta u)(z \circ h)$.

We next consider the terms involving u and f. They are convolutions of Γ, that is a homogeneous kernel of degree $-Q$, with u and f, multiplied by bounded functions compactly supported in $B_\varrho(z_0)$. Moreover, u and f belong to $L^r(B_\varrho(z_0))$, for every r such that $1 \leq r \leq p$. We then choose r such that

$$\frac{1}{r} = \frac{1}{p} + \frac{1}{Q+2}$$

and we apply again Theorem 1.1 with p replaced by r. This concludes the proof. □

Proof of Theorem 1.2 It follows from Proposition 4.2 by a simple covering argument. The constant $\widetilde{\varrho}$ can be chosen as follows. We let

$$\overline{\varrho} := \min\big\{\varrho > 0 \mid B_\varrho(z) \subset \Omega \text{ for every } z \in K\big\},$$

then $\widetilde{\varrho} := \dfrac{\overline{\varrho}}{3c_T}$, so that we can choose $\sigma = \widetilde{\varrho}$ in every ball of the covering of K. □

Proof of Theorem 1.3 If f is a weak solution to (1.22), then it is a weak solution to

$$\partial_t f + \langle v, \nabla_x f\rangle = \Delta_v f + g - \mathrm{div}_v G, \tag{4.3}$$

where $G_j = \partial_{v_j} f$, $j = 1, \dots, n$. Note that the homogeneous dimension of the operator in (4.3) is $Q+2 = n+3n+2$. By our assumptions $G_j \in L^p(\Omega)$ for every $j = 1, \dots, n$. Then the proof can be concluded by the same argument used in the proof of Theorem 1.2. □

5 Conclusion

The method used in this article for Kolmogorov equations can be adapted to the study of a wider family of differential operators, provided that they have a fundamental solution and that are invariant with respect to a suitable Lie group on their domain. Sobolev inequalities for operators of this kind have been proven in [2]. We recall here the assumptions on the operators.

Consider a differential operator in the form

$$\mathscr{L}u := \sum_{i,j=1}^{m} X_j\left(a_{ij}(x,t)X_i u\right) + X_0 u - \partial_t u, \tag{5.1}$$

where $(x,t) = (x_1, \dots, x_N, t)$ denotes the point in $\mathbb{R}^{N+1}$, and $1 \le m \le N$. The X_j's in (5.1) are smooth vector fields acting on $\mathbb{R}^N$, i.e.

$$X_j(x,t) = \sum_{k=1}^{N} b_k^j(x,t)\partial_{x_k}, \qquad j = 0, \dots, m,$$

and every b_k^j is a C^∞ function. In the sequel we always denote by $z = (x,t)$ the point in $\mathbb{R}^{N+1}$, and by A the $m \times m$ matrix $A = \left(a_{i,j}\right)_{i,j=1,\dots,m}$. We also consider the *elliptic* analogous of $\mathscr{L}$

$$\mathscr{L}u := \sum_{i,j=1}^{m} X_j\left(a_{ij}(x)X_i u\right) \tag{5.2}$$

In both cases we assume that the coefficients of the matrix A are bounded measurable functions, and that A is symmetric and uniformly positive, that is, there is a positive constant μ such that

$$\sum_{i,j=1}^{m} a_{ij}(x,t)\xi_i, \xi_j \ge \mu|\xi|^2, \qquad \text{for every } \xi \in \mathbb{R}^m,$$

and for every $(x, t) \in \mathbb{R}^{N+1}$ (or for every $x \in \mathbb{R}^N$ as we consider the operator $\mathscr{L}$ in (5.2)).

Clearly, the Laplace operator Δ and the heat operator $\Delta - \partial_t$ write in the form (5.2) and (5.1), respectively, if we choose $X_j := \partial_{x_j}$ for $j = 1, \ldots, N$, $X_0 := 0$, and the matrix A agrees with the $N \times N$ identity I_N. In the sequel we will use the following notations:

$$X = (X_1, \ldots, X_m), \qquad Y = X_0 - \partial_t, \qquad \mathrm{div}_X F = \sum_{j=1}^{m} X_j F_j,$$

for every vector field $F = (F_1, \ldots, F_m)$, so that the expression $\mathscr{L}u$ reads

$$\mathscr{L}u = \mathrm{div}_X(AXu) + Yu.$$

Finally, when A is the $m \times m$ identity matrix, we will use the notation

$$\mathscr{L}_0 := \sum_{k=1}^{m} X_k^2 + Y.$$

References

1. F. Bouchut, Hypoelliptic regularity in kinetic equations. J. Math. Pures Appl. (9) **81**(11), 1135–1159 (2002)
2. C. Cinti, S. Polidoro, Pointwise local estimates and Gaussian upper bounds for a class of uniformly subelliptic ultraparabolic operators. J. Math. Anal. Appl. **338**, 946–969 (2008)
3. C. Cinti, A. Pascucci, S. Polidoro, Pointwise estimates for solutions to a class of non-homogeneous Kolmogorov equations. Math. Ann. **340**(2), 237–264 (2008)
4. G.B. Folland, Subelliptic estimates and function spaces on nilpotent Lie groups. Ark. Mat. **13**(2), 161–207 (1975)
5. F. Golse, C. Imbert, C. Mouhot, A.F. Vasseur, Harnack inequality for kinetic Fokker-Planck equations with rough coefficients and application to the Landau Equation (to appear on Ann. Sc. Norm. Super. Pisa Cl. Sci. (5). DOI Number: 10.2422/2036-2145.201702_001. Preprint, arXiv.org:1607.08068) (2017)
6. L. Hörmander, Hypoelliptic second order differential equations. Acta Math. **119**, 147–171 (1967)
7. E. Lanconelli, S. Polidoro, On a class of hypoelliptic evolution operators. Rend. Sem. Mat. Univ. Politec. Torino **52**(1), 29–63 (1994)
8. M. Manfredini, S. Polidoro, Interior regularity for weak solutions of ultraparabolic equations in divergence form with discontinuous coefficients. Boll. Unione Mat. Ital. Sez. B Artic. Ric. Mat. **1**(8), 651–675 (1998)
9. A. Pascucci, S. Polidoro, The Mosers iterative method for a class of ultraparabolic equations. Commun. Contemp. Math. **6**, 395–417 (2004)
10. S. Polidoro, M.A. Ragusa, Hölder regularity for solutions of an ultraparabolic equations in divergence form. Potential Anal. **14**, 341–350 (2001)
11. E.M. Stein, Singular integrals and differentiability properties of functions, in *Princeton Mathematical Series*, vol. 30 (Princeton University Press, Princeton, 1970), xiv+290

Mathematical Modeling of the Rod Phototransduction Process

Giovanni Caruso

Abstract Rod photoreceptors are capable of responding to very dim light, thus allowing for night vision. The phototransduction process in rods is a quite well known phenomenon and thus it can be considered as a paradigm of cell signalling mechanisms. Despite a quite deep understanding of the underlying biochemical processes and the availability of measured ranges for all the involved physical parameters, a model able to capture the intricate diffusion processes arising inside the rod cell was not available in the literature. In fact, only simple well stirred models, considering the concentration of the involved species as uniform inside the cell, have been proposed and used by biologists for the interpretation of the experimental measurements. As a matter of fact, due to the intricate cell structure and the photon absorption mechanism, the diffusion processes inside the cell turn out to be crucial for a correct understanding of the cell operation. To this end, a spatially fully resolved model of the rod cell has been proposed, taking into account all the diffusion processes arising inside the rod. The complexity due to the intricate rod cell structure was greatly reduced by using the mathematical techniques of homogenization and concentration of capacity, yielding a model described by coupled partial differential equations of parabolic type, with non-linear source terms, set into simple homogeneous domains. A finite element formulation of the proposed model was then developed, supplying numerical results well reproducing the available experimental results. Once validated, the model was used for performing virtual experiments, not possible in vivo, and testing new hypotheses, providing insights about unknown and debated aspects of the rod phototransduction phenomenon.

Keywords Rod phototransduction · Single photon response · Variability analysis · Three dimensional model · Homogenization · Concentration of capacity

---

G. Caruso (✉)
ITABC-Italian National Research Council, Monterotondo Stazione, Rome, Italy
e-mail: giovanni.caruso@itabc.cnr.it

V. Vespri et al. (eds.), *Harnack Inequalities and Nonlinear Operators*,
Springer INdAM Series 46, https://doi.org/10.1007/978-3-030-73778-8_5

1 Introduction

Retinal rod cells of vertebrates are capable of detecting even a single photon of light, thus allowing for night vision. The cell outer segment has a cylindrical membrane comprising a stack of hundreds of equally spaced and aligned lipidic discs, which are immersed in the rod cytoplasm filling the outer segment. The discs are impermeable to diffusion of species inside the cytosol. A small gap exists between the disc boundaries and the lateral membrane, referred to as outer shell, allowing for vertical diffusion. The phototransduction process is initiated by the absorption of one or more photons by the rhodopsin molecules laying on the discs, originating the disc activation cascade involving the receptor (rhodopsin molecules), the transductin (G proteins) and the effector (phosphodiesterase). These species can diffuse only on the disc surface. Activated effector initiates the cytosol cascade by hydrolyzing the cGMP, locally near the activation location. The local cGMP initial drop, by diffusion through the interdiscal spaces, produces a decrease of cGMP concentration in the outer shell, operating a closure of the cGMP-gated channels located on the lateral membrane. This implies a decrease of the Ca^{2+} ion influx, while the ionic exchangers on the lateral membrane keep on removing Ca^{2+} ions from the cell. This implies a change of the ionic current flowing from the lateral membrane, which constitutes the cell signal. As a consequence, a drop of Ca^{2+} concentration locally on the lateral membrane occurs, then reaching the interdiscal spaces through diffusion. The Ca^{2+} drop stimulates the guanylyl cyclase located on the lipidic discs to produce new cGMP, thus restoring equilibrium conditions and ending the signaling process. This phenomenon can be considered as an archetype of cellular signaling since it is very well understood from an experimental point of view, and reasonable estimates exist for all the physical and biochemistry parameters involved. From a theoretical point of view, a well stirred model describing in detail all the involved processes was developed [18], based on ODE's. In this model the involved quantities, both taking part to the activation cascade and the cytosol cascade, were described in terms of their average concentration in the whole cell, thus neglecting any diffusion process. This model has been widely used by experimentalists to give an interpretation of their measurements. As a matter of facts, the rod single photon response initiates in one point of the surface of one of the discs, producing a local drop of cGMP in the interdiscal space around the activation point. This initial signal must then reach the lateral membrane to produce a change in the ionic current. This occurs through diffusion in the thin interdiscal space adjacent to the activated disc surface, until the outer shell is reached, where vertical diffusion is possible and consequently a spread of the signal. Thus, it appears that diffusion processes are essential for a correct description of the phototransduction phenomenon, and those diffusion processes are quite intricate, since occur on different domains exchanging information to each other. An attempt of taking into account diffusion in the rod cell has been made in very few works

[16, 17], with the development of a radially well stirred model, where diffusion occurs only along the rod axis direction. This simplified model neglects the radial diffusion, which is essential for a correct understanding of the phenomenon and a correct interpretation of the experimental results.

To this end, an interdisciplinary research group coordinated by prof. DiBenedetto was involved into the project of developing a realistic model of the rod phototransduction process, with the goal of developing a software tool implementing a virtual rod for designing virtual experiments, not even possible to be done in practice by experimentalists, and testing new hypotheses. A model describing the diffusion processes arising in the actual intricate rod cell structure is prohibitive, since it requires an enormous computational effort [10] thus not making it suitable for the purposes of implementing a virtual rod helpful for experimentalists. To this end, the homogenization [8, 14, 15, 20] and concentration of capacity [1, 14, 19] mathematical techniques have been used, for alleviating the geometry complexity. The homogenization machinery applied to the rod cell introduces a sequence of diffusion processes parametrized by a small parameter ε, proportional to the thickness of the discs, of the interdiscal layers and of the outer shell. In the homogenization limit $\varepsilon \to 0$, the number of discs and inderdiscal spaces increases and the outer shell thickness becomes infinitesimal, almost disconnecting the inderdiscal spaces from each other. This is a non-standard homogenization problem, requiring specific technicalities described in a series of theoretical papers [2, 4–7]. The resulting homogenized model [3] was described by parabolic equations with non–linear source terms, set in simple homogeneous domains (a cylinder for the inner volume diffusion, a cylindrical surface for the diffusion in the outer shell, and a disc for the diffusion arising in the inderdiscal space above the activated disc surface). Then, a dedicated program in matlab environment was developed, based on a suitable finite element formulation. The numerical results supplied by the homogenized model were compared with results obtained from a model taking into account the actual rod geometry. The comparison between them revealed that the homogenized model was well capable of reproducing all the spatial information provided by the pointwise model [10]. On the other hand, a simulation based on the pointwise model required more than 40 h running on a supercomputer network, whereas an analogous simulation based on the homogenized model took less than 30 s on a standard laptop. Once the homogenized model was validated by comparison with the pointwise model, it has been used for investigating experimental issues and open problems among the research community. The model has the peculiarity of containing only physical parameters whose values can be obtained from available experimental results. With the same consistent set of parameters [21], the model was able to well reproduce available measurements made in rods of both wild–type and transgenic species [12]. Moreover, the model was used for successfully investigating and getting highlights on open issues debated among biologists, like the role of incisures in the photoresponse [11] and the causes of the low variability observed in rod single photon response [13].

2 Mathematical Description of the Rod Phototransduction Process

The rod outer segment (ROS) of vertebrates can be assimilated to a right circular cylinder Ω_ε, of height H and radius $R + \sigma\varepsilon$ for some given, positive parameters R, σ and ε (see in Fig. 1 a micrograph of the salamander ROS). It houses a stack of n parallel, equal, equi-spaced, disc-like, functionally independent bilayers, C_j, $j = 1, 2, \ldots, n$, as depicted in Fig. 2, whose upper and lower faces are denoted

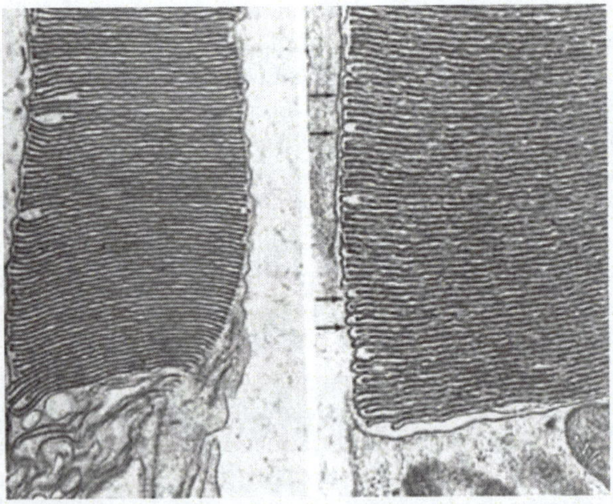

Fig. 1 Micrograph of the salamander ROS

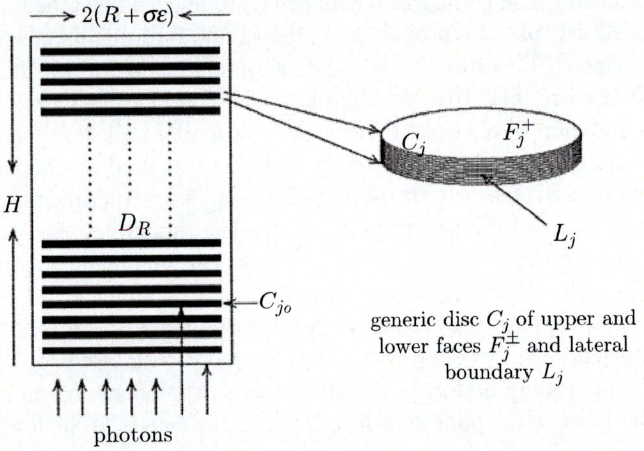

Fig. 2 Cross section of ROS with a stack of discs

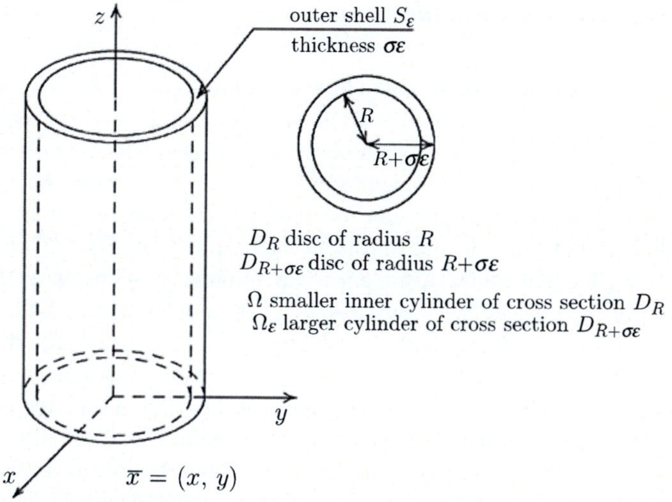

Fig. 3 Schematic representation of a ROS showing the outer shell

with $F_j^{\pm}$, whereas L_j denotes their lateral surface. Each of the bilayers has "radius" R and thickness ε, they are mutually separated by a distance $\nu\varepsilon$. While not discs, we refer to them as discs, following the common biological terminology. The $n+1$ interdiscal spaces are the spaces between two continuous cylinders C_j and C_{j+1}, indicated with I_j, $j = 0 \ldots n$. The outer shell of the ROS, denoted by S_ε is a thin cylindrical shell, of thickness $\sigma\varepsilon$ enclosing the ROS, visible in Fig. 3. The disc number n, and the geometrical parameters H and R, ν, σ, and ε depend on the species. For example, for the Mouse $n \approx 800$, $H \approx 23\,\mu\text{m}$, $R \approx 0.7\,\mu\text{m}$, $\varepsilon \approx 14\,\text{nm}$ and $\nu \approx \sigma \approx 1$. A cartesian coordinate system $(\overline{x}, z) = (x, y, z)$ is introduced, as indicated in Fig. 3.

The phototransduction machinery takes place on two different levels, the first one being the proteinic discs where proteins are embedded, such as rhodopsin (Rh), the light receptor, G protein (G), also called transducin, and cGMP phosphodiesterase (PDE), the effector. These membrane-associated proteins can diffuse only on the face of the disc where they are located (two-dimensional diffusion process). The second level is the three dimensional diffusion of the second messengers calcium (Ca^{2+}) and cGMP (cG) in the fluid cytosol, filling the space between discs and the outer shell. The processes arising at this two levels are described in some detail in the next two sections, where the relevant governing partial differential equations are also reported.

2.1 The Activation Cascade

The signaling process in a rod is initiated when a photon of light hits a molecule of rhodopsin, located on one of the discs, say for example the lower face $F_{j_o}^-$ of the disc C_{j_o} (Fig. 2). The rhodopsin becomes activated (denoted by R^*), after absorbing the photon and, in turn, activates any G protein it interacts with, with certain catalytic rate. Its catalytic rate decreases each time one of its available phosphorylation sites (six for mouse rod) gets phosphorylated by encounter with a rhodopsin kinase molecule. After a certain number of phosphates has been attached, the rhodopsin becomes inactivated by interaction with an arrestin molecule (arresting binding), and consequently its catalytic activity goes to 0. The duration of each phosphorylation state until final quenching by arresting binding is a stochastic variable exponentially distributed, and at each phosphorylation state the possibility of being further phosphorylated or quenched by arrestin is a Bernoulli trial. This situation can be suitably described by a continuous time Markov chain (CTMC), as better specified next. Each of the activated G proteins, denoted with T*, is capable of activating one catalytic subunit of PDE on the disc C_{j_o}, by binding to it upon contact. The bound pair so generated is denoted by E* and is capable of hydrolyzing cGMP molecules giving to the cytosol cascade. Activated E* molecules become inactivated with a certain deactivation rate, thus after rhodopsin quenching by arresting their concentration goes to 0 restoring the initial dark state. This cascade takes place only on one face of the disc C_{j_o}, denoted with D, and is described by the following equations:

$$\frac{\partial \mathrm{T}^*}{\partial t} - D_{\mathrm{T}}\Delta \mathrm{T}^* = \sum_{j=1}^{n} \nu_j \chi_{(t_{j-1},t_j]}(t)\delta_{x_o} - k_{\mathrm{TE}}([\mathrm{PDE}]_o - \mathrm{E}^*)\mathrm{T}^*$$
$$\frac{\partial \mathrm{E}^*}{\partial t} - D_{\mathrm{E}}\Delta \mathrm{E}^* = k_{\mathrm{TE}}([\mathrm{PDE}]_o - \mathrm{E}^*)\mathrm{T}^* - k_{\mathrm{E}}\mathrm{E}^* \tag{2.1}$$

in D, complemented by homogeneous initial data, and no-flux boundary conditions on the boundary of D. Here T*, E* and $[\mathrm{PDE}]_o$ denote, respectively, the surface densities of activated G protein, activated phosphodiesterase and initial concentration of PDE molecules. For low levels of illumination the term $[\mathrm{PDE}]_o - \mathrm{E}^*$ can be well approximated by $[\mathrm{PDE}]_o$. D_{T} and D_{E} indicate, respectively, the diffusion coefficients of T* and E* Moreover, x_o is the location on the disc where photon absorption occurred, k_{TE} is the coupling coefficient from T* to E*. The constant k_{E} is the rate of deactivation of E*. The constant ν_j is the catalytic activity of the rhodopsin molecule R_j^* in its j-th phosphorylation state, i.e. after acquisition of $j-1$ phosphates. The times $s_j = (t_j - t_{j-1})$ are, exponentially distributed, random sojourn times of R^* in its j-th phosphorylation state, with average equal to τ_j. In this model it is assumed that the activated rhodopsin molecule remains

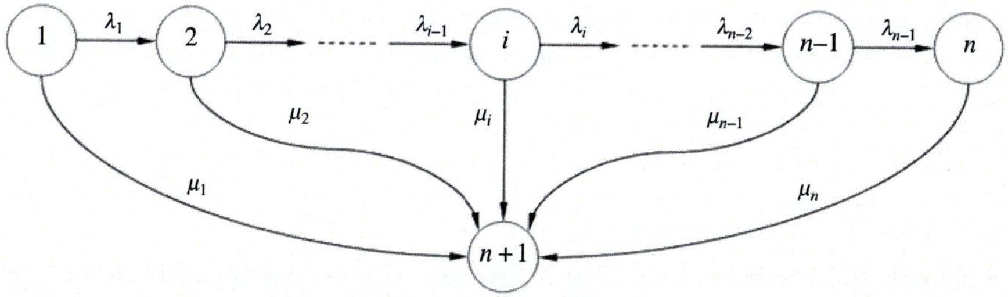

Fig. 4 State diagram of CTMC model for rhodopsin deactivation. States 1 to n are active states and state n+1 is the inactive state. The phosphorylation rates and arrestin binding rates are denoted respectively by λ_i and μ_i

still at x_o during all the activation-deactivation process; as a matter of facts it undergoes a slow random walk motion which is essentially immaterial for the analysis of E* generation [12]. In order to get a deterministic version of (2.1), resembling the average of many and many samples, the stochastic activation term $\sum_{j=1}^{n} \nu_j \chi_{(t_{j-1},t_j]}(t)\delta_{x_o}$ may be replaced by its average over many and many samples

$$\sum_{j=1}^{n} \nu_j P_j(t)\delta_{x_o}\,, \tag{2.2}$$

where P_j indicates the probability of R^* to be in the j-th phosphorylation state at time t. These probabilities can be computed by using the CTMC depicted in Fig. 4, according to the following system of differential equations:

$$\begin{array}{lll} \dot{P}_1 = -(\lambda_1 + \mu_1)P_1, & P_1(0) = 1 & \\ \dot{P}_j = \lambda_{j-1}P_{j-1} - (\lambda_j + \mu_j)P_j, & P_j(0) = 0 & \text{for } j = 2, \ldots, n \\ \dot{P}_{n+1} = \sum_{j=1}^{n} \mu_j P_j, & P_{n+1}(0) = 0\,. & \end{array} \tag{2.3}$$

In (2.3) $n-1$ is the number of available phosphorylation sites, the state $j = 1$ is the newly activated rhodopsin with maximum catalytic activity ν_1, whereas $j = n+1$ is the inactive state upon arresting binding, with $\nu_{n+1} = 0$. The sequence λ_j are the phosphorylation rates whereas μ_j are the arrestin binding rates. For notation consistency $\lambda_n = 0$, since once the last phosphorylation state $j = n$ has been reached, only arresting binding is possible. Biological studies suggest that the sequences ν_j and λ_j can be choosen as

$$\nu = \nu_o e^{-k_v(j-1)}\,, \quad \lambda = \lambda_o(n-j)\,, \tag{2.4}$$

where ν_o, k_v and λ_o can be calibrated through available experimental results [21] together with the arrestin sequence components μ_j. The average τ_j of the sojourn time s_j in the j-th phosphorylation state is given by $\tau_j = 1/(\lambda_j + \mu_j)$. It can be easily checked from (2.3) than the consistency condition $\sum_{j=1}^{n+1} P_j(t) = 1 \quad \forall t \geq 0$ holds. The values of λ_j, μ_j and ν_j depend on the biochemistry of each species and can be calibrated by comparisons with suitable available experimental results.

2.2 *The Cytosol Cascade*

The cytosol cascade occurs in the cytosol filling the interdiscal spaces and the outer shell, where the second messengers cG and Ca^{2+} can freely diffuse. This domain is obtained by removing from the cylinder Ω_ε all the discs C_j, and is denoted with $\tilde{\Omega}_\varepsilon = S_\varepsilon \bigcup_j I_j$. Since within $\tilde{\Omega}_\varepsilon$ there are no volume sources for either cG or Ca^{2+}, the following equations describe their diffusion:

$$\left.\begin{aligned} \frac{\partial \mathrm{cG}}{\partial t} - D_{\mathrm{cG}} \Delta \mathrm{cG} = 0 \\ \frac{\partial \mathrm{Ca}}{\partial t} - D_{\mathrm{Ca}} \Delta \mathrm{Ca} = 0 \end{aligned}\right. \qquad \text{in } \tilde{\Omega}_\varepsilon \tag{2.5}$$

where D_{cG}, D_{Ca} are the respective free diffusivity constants in the cytosol. Active phosphodiesterase E* diffusing on the faces of the activated discs hydrolyzes cGMP diffusing in the cytosol. A small contribution comes from the spontaneously activated phosphodiesterase, not needing photon absorption by rhodopsin molecules, uniformly distributed on the faces of all the discs; a large contribution comes from the E* generated by the light activated rhodopsin on one face of the activated discs (here, the lower face $F_{j_o}^-$ of the disc C_{j_o}). The latter contribution causes a consistent cG drop, initially concentrated in the interdiscal space I_{j_o-1} under the disc C_{j_o}, and then reaching the outer shell and the other interdiscal spaces through diffusion process. When cG drops near the outer shell lateral membrane $\partial S_\varepsilon = \{|\overline{x}| = R + \sigma\varepsilon\} \times [0, H]$, it causes the closure of some of the cGMP-gated channels there located, resulting in a lowering of the influx Ca^{2+} positive ions. Because of the $Na^+/K^+/Ca^{2+}$ exchangers which continue to remove Ca^{2+} from the cytosol, there is a decrease in the calcium concentration, starting from the outer shell and diffusing in all the cytosol, which in turn results in an increase in cG production by stimulation of Ca^{2+}-inhibited guanylyl cyclase, and thus a consequent reopening of the channels. In the meantime, deactivation of the activated rhodopsin by phosphorylation and arrestin binding ends the activation of phosphodiesterase molecules, and thus E* decays to basal, ending depletion of cGMP. These mentioned processes are introduced in the cytosol cascade model as boundary flux terms complementing the field equations (2.5). In particular, on the disc faces $F_j^\pm$ it is set:

$$\mp D_{\mathrm{cG}} \frac{\partial \mathrm{cG}}{\partial z}\bigg|_{F_j^\pm} = \frac{1}{2}\nu\varepsilon \left\{ -\beta_{\mathrm{dark}}\mathrm{cG} + \alpha_{\min} + \frac{\alpha_{\max} - \alpha_{\min}}{1 + (\mathrm{Ca}/\beta)^m} \right\} - \delta_{z_{j_o}} k^* \mathrm{E}^* \mathrm{cG} \tag{2.6}$$

where

$$\begin{array}{lll} \delta_{z_{j_o}} = 1 & \text{if } z = z_{j_o} \text{ (the } z\text{-coordinate of } F_{j_o}^-) \\ 0 & \text{otherwise.} \end{array} \tag{2.7}$$

Furthermore, cG has zero physical flux on each of the disc lateral boundary L_j, on the rod lateral boundary ∂S_ε and on the top ($z = H$) and bottom ($z = 0$) of the rod outer segment. In (2.6) the first term inside the brackets is the hydrolysis contribution operated by the spontaneously activated PDE on all the disc faces, the second term is the cG production operated by guanylyl cyclase and, finally, the last term is the cG hydrolysis operated by the E* arising only on the lower face $F_{j_o}^-$ of the activated disc where photoisomerization occurred. Moreover, β_{dark}, $\alpha_{\min}$, $\alpha_{\max}$, β, m, k^* are positive measurable physical parameters and the term $1/2\nu\epsilon$ is needed to convert volume densities, which are the ones experimentally measured, into surface densities, which are needed for writing the boundary conditions. For more insight on the physics of the phenomenon see [3]. Then the following boundary condition for the Ca^{2+} flux can be written on the lateral boundary $\partial\Omega_\varepsilon$ of the cylinder Ω_ε, where the ionic channels are located:

$$-D_{\text{Ca}}\nabla\text{Ca}\cdot\mathbf{n} = \eta\left(J_{\text{ex}} - \frac{1}{2}f_{\text{Ca}}J_{\text{cG}}\right), \tag{2.8}$$

In (2.8), **n** is the unit normal to the lateral boundary of the rod, pointing outside the rod, η and f_{Ca} are positive physical constants. Moreover

$$J_{\text{ex}} = \frac{j_{\text{ex,sat}}}{\Sigma_{\text{rod}}}\frac{\text{Ca}}{\text{Ca} + K_{\text{ex}}} \tag{2.9}$$

is the current density carried by the calcium ions exiting the rod through the electrogenic exchangers, and

$$J_{\text{cG}} = \frac{j_{\max}}{\Sigma_{\text{rod}}}\frac{\text{cG}^k}{\text{cG}^k + K_{\text{cG}}^k}. \tag{2.10}$$

In (2.9) and (2.10), $j_{\text{ex,sat}}$, Σ_{rod}, K_{ex}, $j_{\max}$, k and K_{cG} are positive measurable physical parameters. Furthermore, the calcium flux is zero across the boundary of each disc C_j and across the bottom ($z = 0$) and the top ($z = H$) of the outer segment. The evolution of the unknowns cG and Ca^{2+} described by the equations (2.5) starts from the dark equilibrium initial conditions

$$\text{cG} = \text{cG}_{\text{dark}}, \quad \text{Ca} = \text{Ca}_{\text{dark}}, \tag{2.11}$$

which can be computed by setting equal to zero the fluxes (2.6) and (2.8), with $\text{E}^* = 0.$ and solving the obtained system of nonlinear equations. The model here described permits one to compute the second messenger concentrations cG and

Ca^{2+} everywhere in the rod. This is much more than the experimentalists can do, since basically they can only measure global quantities such as average volume concentration in rod cells under constant illumination and the total ionic current I flowing across the rod surface as a function of the time, generated by one or more photoisomerizations after a flash delivery. This latter global quantity can be computed, depending on the time evolution of cG and Ca^{2+}, according to the following expression:

$$I(t) = \int_{\partial\Omega_\varepsilon} [J_{\text{ex}}(\overline{x}, z, t) + J_{\text{cG}}(\overline{x}, z, t)]\, dA\,. \tag{2.12}$$

Other two variants of the above quantity, used among experimentalists are:

$$R(t) = \int_{\partial\Omega_\varepsilon} [I_{\text{dark}} - I(t)]\, dA\,, \qquad R_{\text{rel}}(t) = \int_{\partial\Omega_\varepsilon} \frac{R(t)}{I_{\text{dark}}}\, dA\,, \tag{2.13}$$

where I_{dark} is the equilibrium current given by (2.12) evaluated for $t = 0$. $R(t)$ and $R_{\text{rel}}(t)$ measure, respectively, the current suppression and the relative current suppression.

The model equations above described are standard from a mathematical point of view, and possess a unique smooth solution. However, a numerical treatment like finite differences of finite elements must be employed in order to compute the solution. Moreover, the domain where the equations are set is a perforated domain composed of the cylinder Ω_ε where the n discs C_j have been removed, and boundary conditions are enforced on the surfaces of all these discs. This domain is quite complicate since in practical applications n is the order of 1000, and require very intense discretizations implying prohibitive computation time required for getting a solution. This is not acceptable if one has in mind to use the model for designing virtual experiments and for performing parameter sensitivity analysis, both requiring many and many trials. On the other hand, a model which accurately takes into account the complicate diffusion processing arising in the rod cell is essential for a correct interpretation of the experimental results and for theoretically investigating open issues among experimentalists. To this end, a research group coordinated by Prof. Emmanuele DiBenedetto, and composed of mathematicians, engineers, computer scientists and biologists started working on the development of a computationally-efficient three-dimensional diffuse model of the rod signalling process, with the goal of realizing a virtual rod incorporating all the updated known features on its operation, usable for designing virtual experiments and testing new hypotheses about the complex mechanisms at the basis of the phototransduction process.

3 Homogenization and Concentration of Capacity

The repeated disposition of many identical discs inside the rod cylinder suggests the possibility of applying the Homogenization Theory [8, 14, 15, 20] to simplify the pointwise problem (2.5)-(2.6)-(2.8), by performing the homogenization limit $\varepsilon \to 0$, thus letting the disc number n going to infinity, and see what kind of information one might derive out of this limit. The goal is to obtain a model described by partial differential equations set in a much simpler domain than the original one, but capable of reproducing the spatial information supplied by the original pointwise model. In order to simplify the symbolism, the system (2.5) is rewritten as

$$\begin{aligned} \frac{\partial u}{\partial t} - D_u \Delta u = 0 \\ \frac{\partial v}{\partial t} - D_v \Delta v = 0 \end{aligned} \qquad \text{in} \quad \tilde{\Omega}_\varepsilon = \Omega_\varepsilon - \bigcup_{j=1}^{n} \overline{C}_j \tag{3.1}$$

where u and v denote the variables cG and Ca^{2+}. Similarly, the boundary conditions (2.6) is rewritten as:

$$D_u \frac{\partial u}{\partial z}\bigg|_{F_j^{\pm}} = \frac{1}{2} \nu \varepsilon \left\{ \pm \gamma u \mp f_1(v) \right\} - \delta_{z_{jo}} u f_2(x,t) , \quad \text{for } j = 1, 2, \ldots, n , \tag{3.2}$$

where the symbols introduced have an obvious meaning (just compare with (2.6)). On the remaining part of the boundary of $\tilde{\Omega}_\varepsilon$, homogenous flux conditions for u hold. Finally, (2.8) becomes,

$$D_v \nabla v \cdot \mathbf{n} = -g_1(v) + g_2(u) \quad \text{on } \partial S_\varepsilon . \tag{3.3}$$

where the meaning of the introduced symbol can be understand by comparison with (2.8). Homogeneous flux conditions for v hold on the remaining part of the boundary of $\tilde{\Omega}_\varepsilon$. In order to compute the homogenization limit, the problem (3.1)-(3.2)-(3.3) is set in a domain $\tilde{\Omega}_\varepsilon$ where $0 < \varepsilon \le \varepsilon_o$, being ε_o the physical disc thickness, letting ε tending to 0. We denote by u_ε and v_ε the labeled solutions of this set of diffusion problems, called the ε-approximating problems. Since, as a consequence of the homogenization limit, the outer shell S_ε and the interdiscal space shrinks, a suitable rescaling by the factor $\varepsilon_o/\varepsilon$ must be considered in order to compensate for this shrinkage. This is precisely the role of Concentrated Capacity [1, 14, 19]. A sketch of the derivation of the homogenization limit is now illustrated, for a rigorous proof of the homogenization machinery see [2, 4, 6]. Set,

$$a_\varepsilon(x) = \begin{cases} 1 & \text{for } x \in \bigcup_{j \neq j_o - 1} I_j \\ \frac{\varepsilon_o}{\varepsilon} & \text{for } x \in I_{j_o-1} \bigcup S_\varepsilon . \end{cases} \tag{3.4}$$

Consider the family of boundary value problems,

$$\begin{aligned} a_\varepsilon(x)\frac{\partial u_\varepsilon}{\partial t} - D_u \operatorname{div} a_\varepsilon(x)\nabla u_\varepsilon = 0 \\ a_\varepsilon(x)\frac{\partial v_\varepsilon}{\partial t} - D_v \operatorname{div} a_\varepsilon(x)\nabla v_\varepsilon = 0 \end{aligned} \quad \text{weakly in } \tilde{\Omega}_\varepsilon \times (0, T)\,, \tag{3.5}$$

with

$$u_\varepsilon,\ v_\varepsilon \in C\left(0, T; L^2(\tilde{\Omega}_\varepsilon)\right) \bigcap L^2\left(0, T; W^{1,2}(\tilde{\Omega}_\varepsilon)\right) \tag{3.6}$$

and with the boundary and initial data

$$\begin{cases} \frac{\varepsilon_o}{\varepsilon} D_u \frac{\partial u_\varepsilon}{\partial z} = \pm\frac{1}{2}\nu\varepsilon_o\{\gamma u_\varepsilon - f_1(v_\varepsilon)\} - \delta_{z_{jo}} u_\varepsilon f_2(x,t) \\ \qquad\qquad \text{on } F_j^\pm\,,\ j = 1, 2, \ldots, n \\ \nabla u_\varepsilon \cdot \mathbf{n} = 0 \quad \text{on} \begin{cases} L_j\,,\ j = 1, 2, \ldots, n \\ \partial S_\varepsilon \\ z = 0 \text{ and } z = H \end{cases} \\ u_\varepsilon(\cdot, 0) = u_o \qquad \text{in } \Omega_\varepsilon \end{cases} \tag{3.7}$$

$$\begin{cases} \frac{\varepsilon_o}{\varepsilon} D_v \nabla v_\varepsilon \cdot \mathbf{n} = g_2(u_\varepsilon) - g_1(v_\varepsilon) \text{ on } \partial S_\varepsilon \\ \nabla v_\varepsilon \cdot \mathbf{n} = 0 \qquad \text{on} \begin{cases} \partial C_j\,,\ j = 1, 2, \ldots, n \\ z = 0 \text{ and } z = H \end{cases} \\ v_\varepsilon(\cdot, 0) = v_o \qquad \text{in } \Omega_\varepsilon \end{cases}. \tag{3.8}$$

In (3.7) and (3.8), u_o and v_o represent the initial concentrations of cG and Ca^{2+} in the dark adapted equilibrium state (2.11).

3.1 *Computing the Homogenization Limit*

By using the special form of the coefficient $a_\varepsilon(\cdot)$ in (3.4), one checks that when $\varepsilon = \varepsilon_o$ the problem (3.5),(3.7)-(3.8) is exactly (3.1)-(3.2)-(3.3). Next, by a series of formal, heuristic remarks we will attempt to gain an understanding of what kind of limiting equations one might expect from (3.5),(3.7)-(3.8) as $\varepsilon \to 0$.

The Interior Limit

As $\varepsilon \to 0$, the union of all the inderdiscal spaces I_j tends to the cylindrical domain $\Omega = \{|\overline{x}| \le R\} \times (0, H)$. Consider the first of (3.5) written over an interdiscal space I_j different than the activated interdiscal space I_{j_o}. Denote by z_j the z-level of the upper face of the disc C_j. Then it holds

$$\frac{\partial u_\varepsilon}{\partial t} - D_u \Delta u_\varepsilon = 0 \quad \text{in } I_j = \{|\overline{x}| < R\} \times (z_j, z_j + \nu\varepsilon)\,, \tag{3.9}$$

with

$$\begin{aligned} D_u \frac{\partial u_\varepsilon}{\partial z}\bigg|_{z=z_j} &= \tfrac{1}{2}\nu\varepsilon\,\{\gamma u_\varepsilon - f_1(v_\varepsilon)\} \\ D_u \frac{\partial u_\varepsilon}{\partial z}\bigg|_{z=z_j+\nu\varepsilon} &= -\tfrac{1}{2}\nu\varepsilon\,\{\gamma u_\varepsilon - f_1(v_\varepsilon)\}\,. \end{aligned} \tag{3.10}$$

Since $0 < \varepsilon \ll 1$, the domain I_j is thin in the z coordinate. This suggests we replace u_ε and v_ε, with their averages in the z-coordinate within the interdiscal space I_j, i.e.,

$$\begin{aligned} (\overline{x}, t) \to \overline{u}_\varepsilon(\overline{x}, t) &= \frac{1}{\nu\varepsilon} \int_{z_j}^{z_j+\nu\varepsilon} u_\varepsilon(\overline{x}, z, t)dz, \\ (\overline{x}, t) \to \overline{v}_\varepsilon(\overline{x}, t) &= \frac{1}{\nu\varepsilon} \int_{z_j}^{z_j+\nu\varepsilon} v_\varepsilon(\overline{x}, z, t)dz\,. \end{aligned} \tag{3.11}$$

Integrating $(3.5)_1$ in dz over $(z_j, z_j + \nu\varepsilon)$ and dividing by $\nu\varepsilon$ gives, formally

$$\frac{\partial \overline{u}_\varepsilon}{\partial t} - D_u \Delta_{\overline{x}} \overline{u}_\varepsilon = -\frac{1}{2}\{\gamma u_\varepsilon - f_1(v_\varepsilon)\}(\overline{x}, z_j + \nu\varepsilon, t) - \frac{1}{2}\{\gamma u_\varepsilon - f_1(v_\varepsilon)\}(\overline{x}, z_j, t)\,, \tag{3.12}$$

where $\Delta_{\overline{x}}$ indicates the laplacian along only the $\overline{x}$ coordinates. If u_ε and v_ε are smooth, uniformly in ε,

$$\begin{aligned} u_\varepsilon(z_j + \nu\varepsilon) = \overline{u}_\varepsilon + O(\varepsilon)\,,\ u_\varepsilon(z_j) = \overline{u}_\varepsilon + O(\varepsilon), \\ v_\varepsilon(z_j + \nu\varepsilon) = \overline{v}_\varepsilon + O(\varepsilon)\,,\ v_\varepsilon(z_j) = \overline{v}_\varepsilon + O(\varepsilon)\,. \end{aligned} \tag{3.13}$$

Therefore (3.12) takes the approximate form,

$$\frac{\partial \overline{u}_\varepsilon}{\partial t} - D_u \Delta_{\overline{x}} \overline{u}_\varepsilon = -\left[\gamma \overline{u}_\varepsilon - f_1(\overline{v}_\varepsilon)\right] + O(\varepsilon) \quad \text{in } I_j,\ j \neq j_o - 1\,. \tag{3.14}$$

Similar and indeed simpler considerations give

$$\frac{\partial \overline{v}_\varepsilon}{\partial t} - D_v \Delta_{\overline{x}} \overline{v}_\varepsilon = 0 \quad \text{in } I_j,\ j \neq j_o - 1\,. \tag{3.15}$$

In the limit $\varepsilon \to 0$, (3.14) and (3.15) represent a family of diffusion processes parametrized with $z \in (0, H)$, taking place on the disc $|\overline{x}| \leq R$.

Limit on the Activated Disc

As $\varepsilon \to 0$, the activated interdiscal space I_{j_o-1} tends to the disc $D = \{|\overline{x} = R|\}$. To write the first of (3.5) over I_{j_o-1} we take into account the form (3.4) of the coefficients $a_\varepsilon(\cdot)$. To simplify the symbolism also set $i_o = j_o - 1$. Therefore

$$\frac{\varepsilon_o}{\varepsilon}\frac{\partial u_\varepsilon}{\partial t} - \frac{\varepsilon_o}{\varepsilon} D_u \Delta u_\varepsilon = 0 \quad \text{in } I_{i_o} = \{|\overline{x}| < R\} \times \left(z_{i_o}, z_{i_o} + \nu\varepsilon\right), \tag{3.16}$$

$$\begin{aligned} \frac{\varepsilon_o}{\varepsilon} D_u \frac{\partial u_\varepsilon}{\partial z}\bigg|_{z=z_{i_o}} &= \tfrac{1}{2}\nu\varepsilon_o \left\{\gamma u_\varepsilon - f_1(v_\varepsilon)\right\}, \\ \frac{\varepsilon_o}{\varepsilon} D_u \frac{\partial u_\varepsilon}{\partial z}\bigg|_{z=z_{i_o}+\nu\varepsilon} &= -\tfrac{1}{2}\nu\varepsilon_o \left\{\gamma u_\varepsilon - f_1(v_\varepsilon)\right\} \end{aligned} \tag{3.17}$$

Define $\mathring{u}_\varepsilon$ and $\mathring{v}_\varepsilon$ as in (3.11) with averages taken in the interval $(z_{i_o}, z_{i_o} + \nu\varepsilon)$ and integrate (3.16) in dz over $(z_{i_o}, z_{i_o} + \nu\varepsilon)$, getting

$$\frac{\partial \mathring{u}_\varepsilon}{\partial t} - D_u \Delta_{\overline{x}} \mathring{u}_\varepsilon = -\left(\gamma \mathring{u}_\varepsilon - f_1(\mathring{v}_\varepsilon)\right) - \frac{1}{\nu\varepsilon_o} \mathring{u}_\varepsilon f_2(x,t) + O(\varepsilon) \qquad \text{in } I_{i_o}\,. \tag{3.18}$$

Similarly,

$$\frac{\partial \mathring{v}_\varepsilon}{\partial t} - D_v \Delta_{\overline{x}} \mathring{v}_\varepsilon = 0 \qquad \text{in } I_{i_o}\,. \tag{3.19}$$

In the limit $\varepsilon \to 0$, (3.18) and (3.19) describe a diffusion process taking place on D.

Limit in the Outer Shell

As $\varepsilon \to 0$, the outer shell S_ε tends to the cylindrical surface $S = \{|\overline{x} = R|\} \times (0, H)$. To identify the limiting equations in the limiting outer shell write the second of (3.5) in cylindrical coordinates $\rho \in \left(R, R + \sigma\varepsilon\right)$, $\theta \in (0, 2\pi]$, $z \in (0, H)$. Taking into account the form (3.4) of the coefficients $a_\varepsilon(\cdot)$,

$$\frac{\varepsilon_o}{\varepsilon}\frac{\partial v_\varepsilon}{\partial t} - \frac{\varepsilon_o}{\varepsilon} D_v \left\{ v_{\varepsilon,\rho\rho} + \frac{1}{\rho} v_{\varepsilon,\rho} + \frac{1}{\rho^2} v_{\varepsilon,\theta\theta} + v_{\varepsilon,zz} \right\} = 0 \quad \text{in } S_\varepsilon\,. \tag{3.20}$$

The outer shell S_ε is thin in the ρ coordinate. This suggests we replace v_ε and u_ε in S_ε with their ρ-averages, i.e.,

$$\begin{aligned} (\theta, z, t) \to \hat{u}_\varepsilon(\theta, z, t) &= \frac{1}{\sigma\varepsilon} \int_R^{R+\sigma\varepsilon} u_\varepsilon(\theta, \rho, z, t) d\rho\,, \\ (\theta, z, t) \to \hat{v}_\varepsilon(\theta, z, t) &= \frac{1}{\sigma\varepsilon} \int_R^{R+\sigma\varepsilon} v_\varepsilon(\theta, \rho, z, t) d\rho\,. \end{aligned} \tag{3.21}$$

To guess a limiting equation in the outer shell, we assume that the various terms in (3.20) are smooth uniformly in ε. Integrating (3.20) in $d\rho$ over $\big(R, R+\sigma\varepsilon\big)$ and taking into account the flux conditions (3.8) gives

$$\sigma\varepsilon_o\left\{\frac{\partial\hat{v}_\varepsilon}{\partial t} - D_u\left(\frac{1}{R^2}\hat{v}_{\varepsilon,\theta\theta} + \hat{v}_{\varepsilon,zz}\right)\right\} = g_2(\hat{u}_\varepsilon) - g_1(\hat{v}_\varepsilon)$$
$$-\frac{\varepsilon_o}{\varepsilon}D_v\nabla v^{\text{ext}}(R,\theta,z))\cdot\frac{\overline{x}}{R} + \int_R^{R+\sigma\varepsilon}\frac{\varepsilon_o}{\varepsilon}D_v\frac{1}{\rho}\frac{\partial v_\varepsilon}{\partial\rho}d\rho + O(\varepsilon)\,, \tag{3.22}$$

where the apex ext indicates restriction to S_ε. By virtue of (3.8) and the assumed smoothness of $v_{\varepsilon,\rho}$,

$$\int_R^{R+\sigma\varepsilon}\frac{\varepsilon_o}{\varepsilon}D_v\frac{1}{\rho}\frac{\partial v_\varepsilon}{\partial\rho}d\rho = O(\varepsilon)\,. \tag{3.23}$$

Consider now an interdiscal space I_j for $j \neq i_o$ and its adjacent disc C_{j+1}. The lateral boundary of I_j is denoted by Λ_j and the lateral boundary of C_j is denoted by L_j. By the boundary conditions (3.8),

$$D_v\nabla v^{\text{ext}}\cdot\frac{\overline{x}}{R} = \begin{cases} D_v\nabla v^{\text{int}}\cdot\frac{\overline{x}}{R} & \text{in } \Lambda_j;\\ 0 & \text{in } L_j\,,\end{cases} \tag{3.24}$$

The jump from Λ_j to L_j suggests we approximate ∇v^{ext} by its average over the interval $\big(z_j, z_j+(1+\nu)\varepsilon\big)$, getting

$$\frac{\varepsilon_o}{\varepsilon}D_v\nabla v^{\text{ext}}\cdot\frac{\overline{x}}{R} = \frac{1}{(1+\nu)\varepsilon}\int_{z_j}^{z_j+(1+\nu)\varepsilon}\frac{\varepsilon_o}{\varepsilon}D_v\nabla v^{\text{ext}}\cdot\frac{\overline{x}}{R}dz$$
$$= D_v\frac{1}{(1+\nu)\varepsilon}\int_{z_j}^{z_j+\nu\varepsilon}\nabla v^{\text{int}}\cdot\frac{\overline{x}}{R}dz = (1-\theta_o)D_v\left.\frac{\partial\overline{v}_\varepsilon}{\partial\rho}\right|_{|\overline{x}|=R} + O(\varepsilon) \tag{3.25}$$

where $\theta_o = 1/(1+\nu)$ and $\overline{v}_\varepsilon$ is defined in (3.11). Similarly for $j = i_o$

$$\frac{\varepsilon_o}{\varepsilon}D_v\int_{z_{i_o}}^{z_{i_o}+\nu\varepsilon}\nabla v^{\text{ext}}\cdot\frac{\overline{x}}{R}\Big|_{\Lambda_{i_o}}dz = \nu\varepsilon_o\left.\frac{\partial\mathring{v}_\varepsilon}{\partial\rho}\right|_{|\overline{x}|=R} + O(\varepsilon)\,, \tag{3.26}$$

Thus (3.22) becomes

$$\frac{\partial\hat{v}_\varepsilon}{\partial t} - D_v\Delta_S\hat{v}_\varepsilon = -\frac{(1-\theta_o)D_v}{\sigma\varepsilon_o}\overline{v}_{\varepsilon,\rho}\big|_{|\overline{x}|=R} - \delta_{z_{jo}}\frac{\nu D_v}{\sigma}\mathring{v}_{\varepsilon,\rho}\big|_{|\overline{x}|=R}$$
$$+\frac{1}{\sigma\varepsilon_o}\big\{g_2(\hat{u}_\varepsilon) - g_1(\hat{v}_\varepsilon)\big\} + O(\varepsilon) \qquad \text{in } S_\varepsilon\,, \tag{3.27}$$

where

$$\Delta_S \hat{v}_\varepsilon = \left(\frac{1}{R^2} \frac{\partial^2 \hat{v}_\varepsilon}{\partial \theta^2} + \frac{\partial^2 \hat{v}_\varepsilon}{\partial z^2} \right) \tag{3.28}$$

is the Laplace-Beltrami operator on S. By similar and simpler considerations,

$$\frac{\partial \hat{u}_\varepsilon}{\partial t} - D_v \Delta_S \hat{u}_\varepsilon = -\frac{(1-\theta_o) D_v}{\sigma \varepsilon_o} \overline{u}_{\varepsilon,\rho} \big|_{|\overline{x}|=R} - \delta_{z_{jo}} \frac{\nu D_v}{\sigma} \mathring{u}_{\varepsilon,\rho} \big|_{|\overline{x}|=R} + O(\varepsilon) \qquad \text{in } S_\varepsilon\,. \tag{3.29}$$

Passing to the limit $\varepsilon \to 0$, (3.29) and (3.27) describe a diffusion process taking place on the surface S. In particular, it clearly appears that the flux coming from the boundary of Ω and of D, rescaled by suitable factors depending on the rod geometrical parameters, are source terms for the diffusion in the cytosol together with the fluxes g_1 and g_2 entering the rod lateral boundary; this is the result of the concentration of capacity of the outer shell and the activated interdiscal space.

3.2 Variational Formulation

The homogenized model previously described contains three pair of unknowns, namely $(\overline{u}, \overline{v})$ defined in the interior Ω, $(\mathring{u}, \mathring{v})$ defined in the special disc D and $(\hat{u}, \hat{v})$ defined in the outer shell surface S. It can be formulated in a weak form, which is useful for a rigorous mathematical treatment and for the derivation of a finite element formulation, as described in the next section. A weak form of the homogenized model can be derived by multiplying by a test function (3.5), integrating by parts taking into account the boundary conditions (3.7) and (3.7), suitably extending u_ε to all the domain Ω and finally taking the limit $\varepsilon \to 0$. All these passages are rigorously detailed in [2, 4–6]. By indicating with $\cdot_T$ the spatio-temporal domain $\cdot \times (0, T)$, the obtained weak formulation is written as

$$\begin{aligned}
&(1-\theta_o) \left\{ \int\!\!\int_{\Omega_T} \{ \overline{u}_t \varphi + D_u \nabla_{\overline{x}} u \cdot \nabla_{\overline{x}} \varphi \}\, dx dt + \int\!\!\int_{\Omega_T} (\gamma \overline{u} - f_1(\overline{v})) \varphi \, dx dt \right\}_{\text{interior}} \\
&\quad + \sigma \varepsilon_o \left\{ \int\!\!\int_{S_T} \{ \hat{u}_t \varphi + D_u \nabla_S \hat{u} \cdot \nabla_S \varphi \}\, dS dt \right\}_{\text{outer shell}} \\
&\quad + \nu \varepsilon_o \left\{ \int\!\!\int_{D_T} \{ \mathring{u}_t \varphi + D_u \nabla_{\overline{x}} \mathring{u} \cdot \nabla_{\overline{x}} \varphi \}\, d\overline{x} dt + \int\!\!\int_{D_T} \{ \gamma \mathring{u} - f_1(\mathring{v}) \} \varphi \, d\overline{x} dt \right. \\
&\qquad \left. + \frac{1}{\nu \varepsilon_o} \int\!\!\int_{D_T} \mathring{u} f_2(\overline{x}, t) \varphi \, d\overline{x} dt \right\}_{\text{activated level}} = 0
\end{aligned} \tag{3.30}$$

for all testing functions $\varphi \in C^1(\overline{\Omega}_T)$ vanishing for $t = T$;

$$(1-\theta_o)\left\{\int\int_{\Omega_T} \{\overline{v}_t\psi + D_v\nabla_{\overline{x}}\overline{v}\cdot\nabla_{\overline{x}}\psi\}\,dxdt\right\}_{\text{interior}}$$

$$+\sigma\varepsilon_o\left\{\int\int_{S_T} \{\hat{v}_t\psi + D_v\nabla_S\hat{v}\cdot\nabla_S\psi\}\,d\eta dt + \frac{1}{\sigma\varepsilon_o}\int\int_{S_T}\{g_1(\hat{v}) - g_2(\hat{u})\}\psi\,d\eta dt\right\}_{\text{outer shell}}$$

$$+\nu\varepsilon_o\left\{\int\int_{D_T} \{\mathring{v}_t\psi + D_v\nabla_{\overline{x}}\mathring{v}\cdot\nabla_{\overline{x}}\psi\}\,d\overline{x}dt\right\}_{\text{activated level}} \tag{3.31}$$

for all testing functions $\psi \in C^1(\overline{\Omega}_T)$ vanishing for $t = T$. The functions $(\overline{u}, \mathring{u}, \hat{u})$ and $(\overline{v}, \mathring{v}, \hat{v})$ are in the following regularity classes

$$\begin{aligned}
&\overline{u}, \overline{v} \in C\big(0, T; L^2(\Omega)\big); \ |\nabla_{\overline{x}}u|, \ |\nabla_{\overline{x}}v| \in L^2(\Omega_T);\\
&\mathring{u}, \mathring{v} \in C\big(0, T; L^2(D)\big); \ |\nabla_{\overline{x}}\mathring{u}|, \ |\nabla_{\overline{x}}\mathring{v}| \in L^2(D_T);\\
&\hat{u}, \hat{v} \in C\big(0, T; L^2(S)\big); \ \big|(\hat{u}_z, \hat{u}_\theta)\big|, \ \big|(\hat{v}_z, \hat{v}_\theta)\big| \in L^2(S_T).
\end{aligned} \tag{3.32}$$

Moreover the following relations hold:

$$\begin{aligned}
&\hat{u}(\theta, z, t) = \overline{u}(\overline{x}, z, t)\big|_{|\overline{x}|=R} \ \text{ in } L^2\big((0, 2\pi] \times (0, T]\big) \quad \text{for all } z \neq z_{j_o};\\
&\hat{u}(\theta, z_{j_o}, t) = \mathring{u}(\overline{x}, t)\big|_{|\overline{x}=R} \ \text{ in } L^2\big((0, 2\pi] \times (0, T]\big);\\
&\hat{v}(\theta, z, t) = \overline{v}(\overline{x}, z, t)\big|_{|\overline{x}|=R} \ \text{ in } L^2\big((0, 2\pi] \times (0, T]\big) \text{ for all } z \in (0, H).
\end{aligned} \tag{3.33}$$

It is readily seen that (3.30)-(3.31) is the weak formulation of (3.5)-(3.7)-(3.8) as $\varepsilon \to 0$. Conversely sufficiently smooth solutions of (3.30)-(3.31) admit the pointwise formulation obtained from (3.5)-(3.7)-(3.8) as $\varepsilon \to 0$. The formulation (3.30)-(3.31) is quite general; for example it can accounts for multiple photon activation, with photons hitting different discs at different z-levels and in different points $\overline{x}$ different for each special disc. It can be used for predicting either deterministic or stochastic phototransduction events, depending on the kind of disc activation chosen as described in Sect. 2.1. Finally, the weak formulation (3.30)-(3.31) has been generalized in [7], in order to account also for incisures, i.e. aligned thin vertical cuts in the rod discs, allowing direct connection between all the interdiscal spaces (other than the outer shell annular space). Such domains does not contain sources and undergoes a thickness shrinkage during the homogenization limit, thus appearing in the weak formulation as a diffusion process described by the Laplace-Beltrami operator in the incisure plane (see [7, 11] for details). The rods of some species like amphibians bear a high number of long incisures, which have several complex functions in addition to the obvious one of increasing and making faster the photoreceptor response, as shown in the foregoing.

4 A Finite Element Formulation

The variational formulation of the homogenized model (3.30)-(3.31) is set in a much simpler domain than the original pointwise model described in Sect. 2, which was set in $\tilde{\Omega}_\varepsilon$, containing the stack of hundreds of holes and relevant boundary conditions. A finite element formulation of the homogenized model will require the discretization of only the three simple domains Ω, D and S, i.e. a cylinder, a disc and a cylindrical surface. To this end, the disc D has been discretized into three-node triangles, the cylinder Ω into prisms with triangular bases, obtained as cartesian product of the disc D triangle discretization with z. Finally, the outer shell S has been discretized into rectangles which coincide with the prism faces laying on the outer shell. These domain discretizations, for a salamander rod geometry, are represented in Figs. 5, 6, and 7. In particular, 23 incisures are present, consisting of additional rectangular surfaces discretized with rectangles as shown in Fig. 7.

According to the finite element formulation, the spatial integrals in (3.30) and (3.31) are numerically evaluated by interpolating the unknowns u and v inside each element by using the standard isoparametric mapping [22] and suitable shape functions. In particular, linear shape functions are used for interpolation in the triangles, and bilinear shape functions for interpolation in the rectangles and prisms. A finite-difference scheme named *Wilson-theta method*, which requires an iterative procedure due to the presence of nonlinear forcing terms in the model, is adopted for the time integration. The formulation is implemented into a dedicated program in matlab environment in a very general manner, allowing for either deterministic or stochastic simulations, single or multiple activation, accounting for the presence of incisures of any number and size, and able to test different hypotheses concerning activation biochemistry and cytosol cascade components. Thus it is very suitable for designing virtual experiments which give responses in a very brief time and are very helpful to biologists for the assessment of the reliability of certain experimental

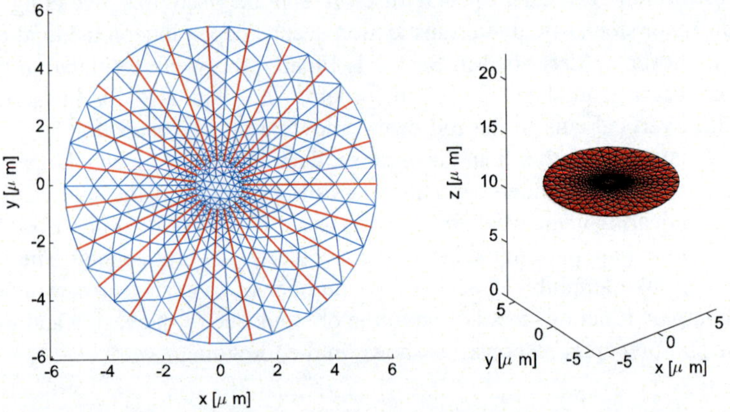

Fig. 5 Triangle discretization of the disc D, including 23 incisures

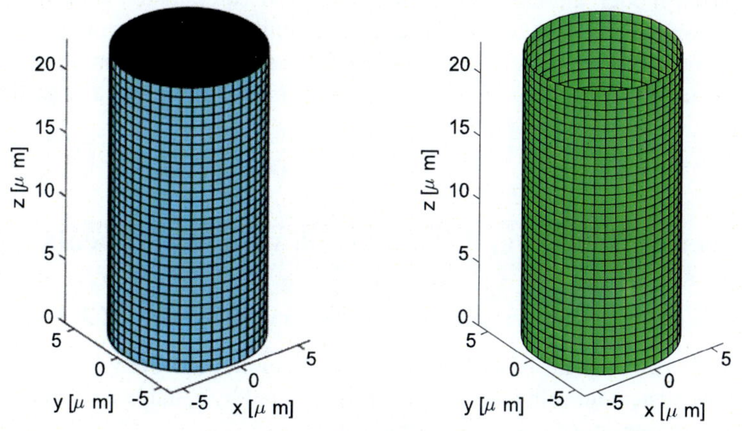

Fig. 6 Prism discretization of Ω (on the left) and rectangle discretization of S (on the right)

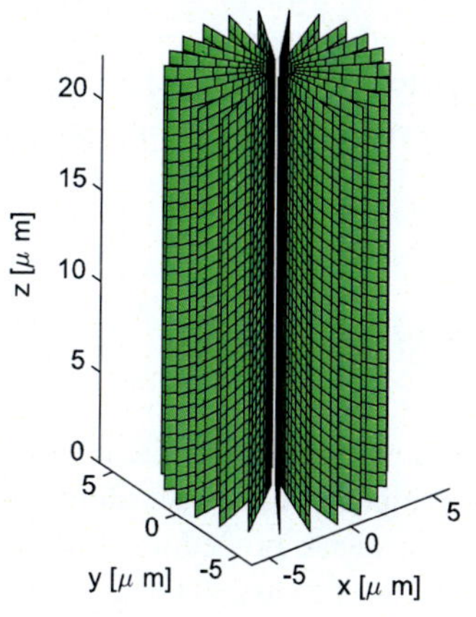

Fig. 7 Rectangle discretization of the 23 incisures

campaigns or for testing new hypotheses. In a first phase, the finite element program based on the homogenized model has been validated with respect to results provided by a finite-difference solution of the original pointwise formulation described in Sect. 2, in order to assess the capability of the former to reproduce the spatially-resolved features arising in the real complex geometry of the rod cell. The results of this investigation can be found in detail in [10]; they confirmed that the homogenization model is able to reproduce all the essential spatial information supplied by the original model set in the actual complex rod geometry. On the other hand, a simulation to compute the single photon response based on the original non-homogenized model took almost 40 h, running on a supercomputer facility, whereas less than 30 s were needed for running a simulation on a standard laptop using the homogenized model, amounting at almost a 1/5200 computation time reduction. This comparative study validated the proposed finite element formulation of the homogenized model, thus resulting suitable for the implementation of a "virtual rod", to be used by experimentalists and researchers within the visual-transduction community.

5 Numerical Results and Comparison with Experiments

A mathematical model of a certain physical phenomenon is successful if

- it is rather "simple", i.e. described by simple equations containing few parameters of clear physical meaning and that can be measured by suitable experiments;
- it is able to reproduce known experimental results;
- it is helpful in explaining unknown and still debated aspects of the phenomenon under investigation.

In the foregoing, some of the main results obtained by using the proposed homogenized model are illustrated, demonstrating the model "success". The details of the biological issues underlying the numerical simulations we show and their significance will be briefly discussed, suggesting relevant publications for getting further insight.

5.1 Reproducing SPR Measurements

As previously highlighted, rod photoreceptors are capable of responding to even a single photon, thus favouring night vision. Accordingly, experimentalists involved in this research developed sophisticate procedures for measuring the single photon response (SPR) of a rod cell. As discussed later, these responses are not reproducible, since the activation-deactivation mechanisms arising in the disc cascade are of stochastic nature. In fact, a big issue which has been studied with the homogenized model is the variability of the SPR. Moreover, each single measurement is quite noisy. In order to extract general information from the measured SPR's, each single measure is suitably normalized and an average is finally performed over many and many trials, to get a SPR representative of the investigated species. In order to reproduce numerically the measured representative SPR's with the homogenized model, an average of many and many stochastic numerical trials should be performed, since the model governing equations contain nonlinear terms. However, for the scope of reproducing the representative experimental measurements, it is acceptable to compute a deterministic numerical simulation by replacing the stochastic activation cascade with its average over many and many samples, given by the solution of the CMTC (Fig. 4) as described in Sect. 2.1. In Fig. 8 the SPR of a salamander and mouse rod are reported, together with the corresponding numerical

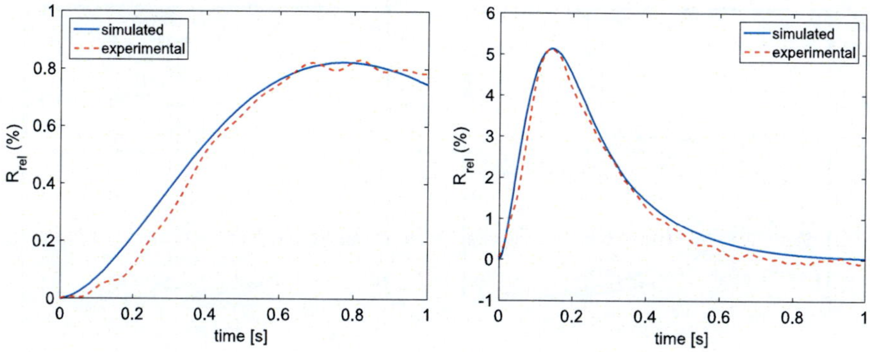

Fig. 8 Experimental single photon responses in salamander (on the left) and in mouse (on the right). Comparison with simulated responses according to the homogenized model

simulations obtained by using the homogenized model. Both the simulations have been obtained by choosing the physical parameters appearing in the model inside the experimental ranges available in the literature for the two species. The biochemical sequences λ_j, μ_j and ν_j have been calibrated according to the most updated and accepted hypotheses among the research community [21].

As shown in Fig. 8, the SPR originates from the dark equilibrium ($R_{\mathrm{rel}} = 0$), and rise up during the activation phase, due to the closure of the cG gated channels implying a lowering of the current I and thus an increase of the relative response R_{rel}. The peak is reached quite after the rhodopsin molecule has been quenched; the delay between peak of E* and peak of R_{rel} is due essentially to the diffusion processes and calcium feedback mechanism operated by the cyclase. After the peak has been reached, the relative response decreases with a final slope depending essentially on k_{E}, finally stabilizing to 0 after initial dark conditions have been restored. The time to peak and peak value change, depending on the considered species. Figure 8 shows that the mouse SPR is much faster and intense than the salamander SPR; this is due both to the different biochemical parameters and the different rod size (salamander has a rod with cross section much larger than mouse, with R=5.5 μm for salamander and 0.7 μm for mouse). The homogenized model is able to well reproduce the experimental results relevant to both the considered species. Biologists are able to make measurements not only on "wild-type" species, but can also create suitable mutants by operating genetic manipulations with the goal of obtaining "mutant" rods where some characteristics have been modified from the wild-type rod. These mutants are referred to as knockout (KO) transgenic mice, since some target characteristics of their rod cells have been switched off (i.e. not expressed, knocked out) through gene manipulation. This is used to experimentally investigate the role played by a certain characteristic in the phototransduction phenomenon. Thus, the predictive capabilities of the homogenized model have been tested also against the available mutant SPR's, as depicted in Figs. 9 and 10 where several mouse mutants have been considered. In particular, with suitable

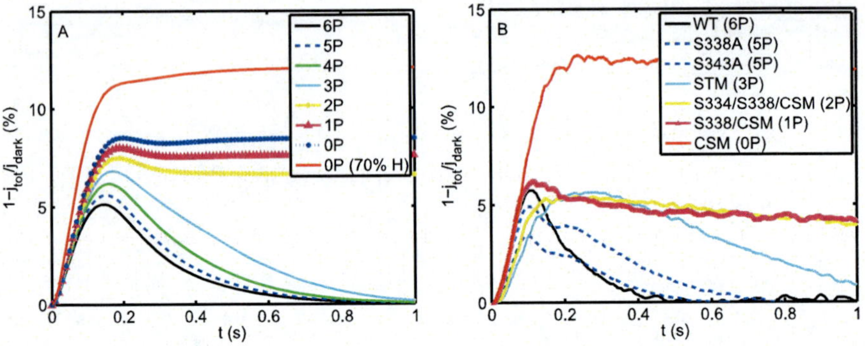

Fig. 9 Simulated (on the left) and experimental (on the right) single photon responses in wild type and mutant mouse rods

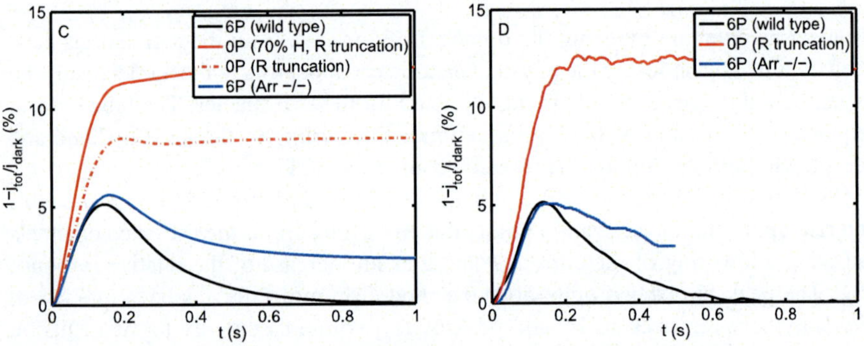

Fig. 10 Simulated (on the left) and experimental (on the right) single photon responses in wild type and mutant mouse rods

gene manipulations, it was possible to grow mouses with rods exhibiting rhodopsin with a modified number of available phosphorylation sites (indicated in the figure legends with the number of sites followed by P), (P = 6 in wild type rods). This amounts at slowing down the recovery mechanism, thus increasing the response peaks and prolonging the response tails. Also in Fig. 10 the mutant indicated with Arr-/- has rods without arrestin, and thus the activated rhodopsin undergoes all the phosphorylation steps and, once six phosphates have been attached to it, remains in this state without being quenched. In this case the experimental SPR has a tail with an horizontal asymptote at an height one half of the peak response. This is a very helpful characteristic to be used for calibrating the biochemical parameters of the disc cascade in the model, together with the activated rhodopsin average lifetime which can also be experimentally determined (for a detailed discussion on how choosing the model parameters according to the available experimental findings see [21]).

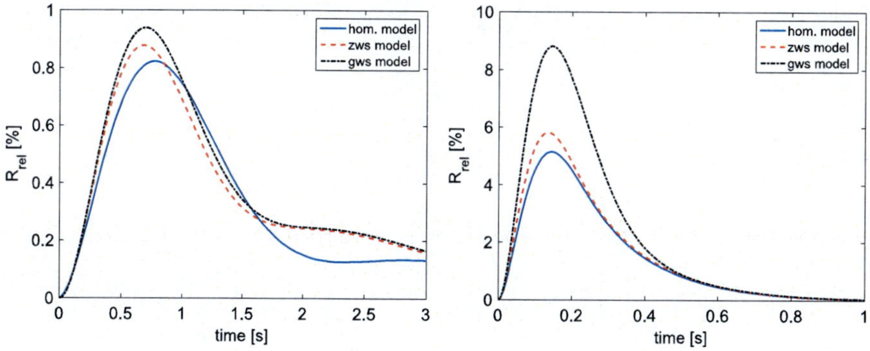

Fig. 11 Comparison of simulated SPR predicted by the homogenized, the radially well stirred and the globally well stirred models. Salamander (on the left), mouse (on the right)

The results confirm the ability of the homogenized model in predicting the responses relevant also to the considered set of mutant rods. It is here remarked that all the above numerical SPR simulations have been obtained by using the same set of parameters as the one used for the wild type response, and varying only the number n of available phosphorylation sites for computing the disc cascade sequences λ_j, μ_j, ν_j depending on the considered mutant, or setting $\mu_j = 0\ \forall j \in [0, n]$ in the case of the Arr-/- mutant (no possibility for activated rhodopsin to be quenched). Finally, in Fig. 11 a comparison is shown between SPR's computed with the homogenized model and with well stirred models existing in the literature and commonly used by experimentalists. In particular, the zws model [16, 17] is a model considering diffusion only along the z rod axis direction and averaging concentrations along the radial direction, whereas the gws is a totally lumped model [18], considering concentrations averaged on all the rod cell. It clearly appears that the well stirred models predict a smaller time to peak and overestimate the peak response, the gws model being obviously less accurate than the zws model.

The simulations in Fig. 12, relevant to a salamander rod, are indeed intended to show the local nature of the SPR. In particular, on the left panel of the figure the current density drop is evaluated on the outer shell along an axis parallel to z at various values of t, where $J(z, \theta, t) = J_{ex} + J_{cG}$ is the current surface density appearing inside the integral in (2.12). Clearly, the response is concentrated on a small portion of the cylindrical surface around the activated level (the disc at the rod midspan). On the right panel, the z profiles of the cG drop are represented, where the cG drop is defined as

$$1 - \frac{\mathrm{cG}(\overline{x}, z, t)}{\mathrm{cG}_{\mathrm{dark}}}. \tag{5.1}$$

In particular, the cG drop is computed along the rod axis $\overline{x} = (0, 0)$, and along an axis parallel to the rod axis and laying on the outer shell (the angular coordinate is almost immaterial for the symmetry of the considered case), in correspondence of

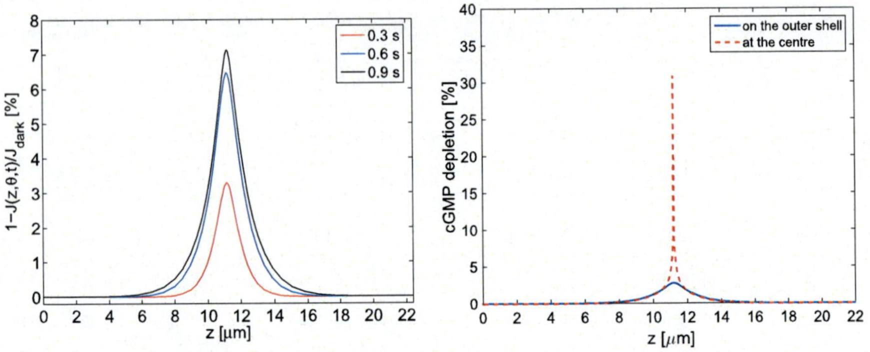

Fig. 12 Local nature of the SPR in a salamander rod. Activation occurring at the disc centre. On the left: profile along z of the current density drop at different values of t; On the right: profiles along z of the cG concentration taken on the rod axis and on an axis laying on the outer shell, at the time to peak

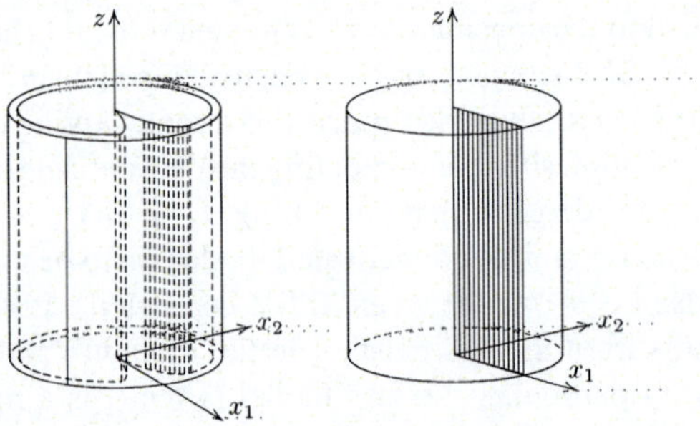

Fig. 13 Schematic representation of a single incisure (mouse case) with its real geometry (on the left), and after concentration of capacity has been applied (on the right)

the time to peak. The drop differences between the disc centre and the outer shell is clearly dramatic in the rod portion around the activated level, highlighting a strong radial gradient of cG concentration.

5.2 *Investigating of the Role of Incisures*

Incisures are vertical aligned cuts through all the rod discs, creating vertical channels contributing to diffusion of second messengers along the rod axis direction, together with the outer shell space. These incisures appears in the rods of most of the species, with different size and number. A schematic representation of a single incisure is depicted in Fig. 13. As mentioned in Sect. 3, incisures are easily accounted for in the homogenized model by applying to them the concentration of capacity; each incisure becomes a rectangular domain as depicted on the right part of Fig. 13 where

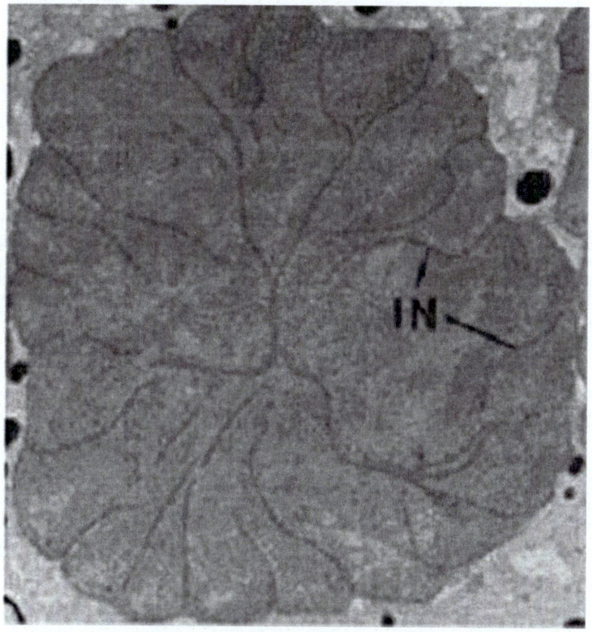

Fig. 14 Micrograph showing the cross section of a frog rod. Several long and thin incisures are clearly visible

diffusion of second messengers occurs, without internal sources. As a matter of fact the role of incisures and the way they affect the rod photoresponse is quite complex and debated among researchers. Several amphibians, like frog and salamanders, have rods bearing many long and thin incisures, see for example Fig. 14 showing a micrograph of a frog rod cross section where incisures are clearly visible. In addition to the obvious function of enhancing second messengers diffusion along the rod, thus increasing the receptor response to light absorption, they have several other functions [11]. For example, the incisures shown in Fig. 14 play a role also in the activation-deactivation disc cascade, since they are barriers for the diffusion processes arising in the disc. More specifically, they create lobes in the disc, thus confining the activated E* inside the lobe where the light photon has been absorbed. In order to analyze all these issues a spatially-resolved model, able to account for the actual spatial disposition of the incisures and for all the diffusion processes arising at the various levels in the cell, is mandatory [11].

Results in Fig. 15 show the influence of the number of incisures (of all identical size) on the rod SPR. In particular, in the left panel a salamander rod is considered, bearing 23 long and thin incisures. The SPR is computed for the wild-type rod, and for rods identical to the wild-type one but with a decreased number of incisures. The results clearly highlight the role of incisures in increasing the SPR, by increasing the overall cross section area available for second messengers diffusion along the z axis. In fact, the increased diffusion allows for an wider region of the lateral

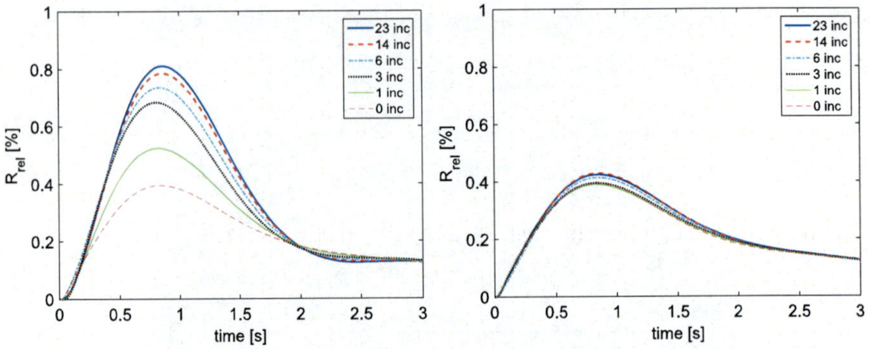

Fig. 15 SPR in salamander with activation at the centre of the midspan disc, in correspondence of different numbers of equally spaced and identical incisures. On the left: long and thin incisures (wild-type case); On the right: incisures with area equal to the previous case, but half long and double thick

membrane where the cG drops and the cG-gated channels close, thus producing a larger response. However, the situation is quite more complex than that. In fact, on the right panel of Fig. 15 a similar analysis is presented, but with incisures of the same area of the wild-type incisures considered in the left panel, but with half length and double thickness. As it clearly appears, the influence of the incisures on the SPR peak is quite reduced than in the previous case, though the area of each incisure has not been varied. This is because in this latter case the incisures are much more distant from the activation site, at the centre, where the activated E* is concentrated and operates the cG hydrolysis. It is here remarked that a radially well stirred model can account for the presence of incisures by only modifying the equivalent diffusion coefficient along z as a function of the overall cross sectional area available for vertical diffusion [16, 17], thus not distinguishing between the two cases in Fig. 15. Only a fully diffuse model can take into account the actual incisure distribution with respect to the activation position, which greatly affects the rod response.

A further investigation regarding the role of incisures, possible only using a fully spatially resolved model, is presented in Fig. 16, where the SPR in a salamander rod is reported in correspondence of different activation locations on the midspan disc. In particular, on the right panel the case of a rod where incisures have been virtually removed is considered. It clearly appears that the SPR peak increases with the increase of the distance of the activation position from the disc centre, and reaches its maximum when the photon hits the disc in correspondence of its boundary. This is because the closer the photoisomerization to the disc boundary, the larger the cG drop in correspondence of the lateral membrane where the cG-gated channels are located. On the other hand, if incisures are present (see panel on the left), the SPR corresponding to different activation locations are quite similar since incisures enhance diffusion process of cG, compensating for the differences in

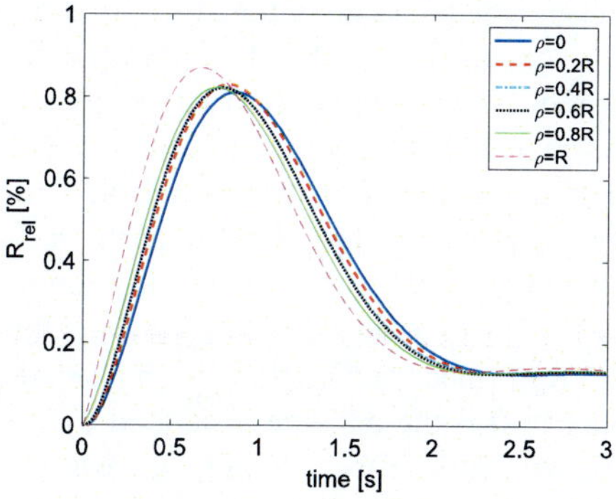

Fig. 16 SPR in salamander for different activation locations. On the left: wild-type salamander rod exhibiting 23 incisures; On the right: same rod but without incisures

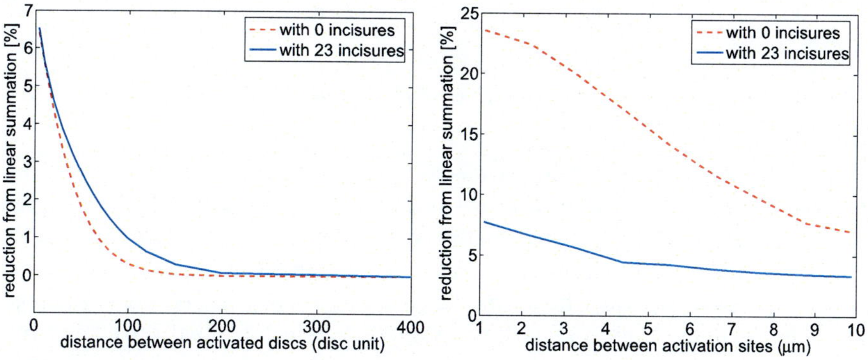

Fig. 17 Reduction from linear summation for response relevant to the simultaneous absorption of two separate photons in a salamander rod, versus the photon separation distance, considering or not the presence of incisures. On the left: the photons are absorbed by two discs at their centre; On the right: the photons are absorbed by same disc, at the rod midspan, on a diameter and symmetrically with respect to the disc centre

photoisomerization position. This fact is crucial for decreasing the SPR variability depending on the activation position in species exhibiting rods of large cross section.

Another important issue regarding rods is the capability of acting as a photon counter, i.e. to respond proportionally to the number of photoisomerizations occurred during a certain small time interval (light flash absorption). This is important for acuity in night vision. To this end, in Fig. 17 the reduction from linear summation relevant to the response of a rod excited by two absorbed photons is reported, versus the distance between the two photoisomerizations. Reduction from linear summation is defined as

$$\frac{R(1) + R(2) - R(1+2)}{R(1+2)}, \tag{5.2}$$

where $R(1)$ and $R(2)$ indicate the rod response corresponding to the absorption of only the photon 1 and photon 2, respectively, whereas $R(1+2)$ is the rod response corresponding to the simultaneous absorption of both the photons. Two situations are considered: on the left panel the photons are absorbed in correspondence of the centre of two separated discs (the disc separation is reported on the abscissa in disc units); on the right panel the two photons are both absorbed by the disc at the rod midspan, along a diameter and symmetrically with respect to the centre (the photon distance is reported on the abscissa).

The simulations show that the reduction from linear summation is very small for photon absorbed at different levels, going to 0 if the disc are separated by at least 200 discs. In this case the incisures have a little effect, slightly increasing the reduction from linear summation since they increase the diffusion along z thus favouring an overlapping of the response spread relevant to the two photons. On the other hand, reduction from linear summation for photons on the same disc is quite large if incisures are absent, and is greatly decreased by the presence of incisures.

5.3 *Investigating the Variability of the SPR*

The simulations presented in the previous sections have been performed on the basis of a deterministic setting, as previously explained. Thus, they resemble the average of many and many stochastic trials. In this section the stochastic nature of the SPR response is analyzed. There are many randomness sources in the SPR machinery [9, 12], the most important being the random times when the phosphorylations occur, the random number of phosphorylation steps before final quenching by arrestin encounter, and the random time at which the final quenching happens. Furthermore, an important component of randomness is also the random location of the photoisomerization on the activated rod disc, the z level of the activated disc being immaterial due to the rod cylindrical structure. These randomness sources introduce a variability in the surface density E^* of the effector, and thus in its total mass $\mathrm{E}^*_{\mathrm{tot}}$ (integral of the E^* over the disc). This variability can be measured in several ways; a statistical representative measure is the coefficient of variation CV, defined as the standard deviation divided by the mean, of some random variable. This measure is widely employed by experimentalists studying the variability of the SPR, applied to the quantities

$$\mathrm{E}^*_{\mathrm{tot,area}}(t) = \int_0^t \mathrm{E}^*_{\mathrm{tot}}(\tau)\, d\tau\,, \qquad R_{\mathrm{rel,area}}(t) = \int_0^t R_{\mathrm{rel}}(\tau)\, d\tau\,, \tag{5.3}$$

which are representative, for values of t high enough, of all the SPR time course, i.e. the rising phase, the peak value and the tail. Based on the system (2.1), the CV of $\mathrm{E}^*_{\mathrm{tot,area}}$ can be analytically computed in terms of the biochemistry sequences ν_j (activated rhodopsin catalytic activity in the j-th phosphorylation state) and τ_j (average sojourn time in the j-th phosphorylation state), under the simplifying assumption that rhodopsin final quenching by arresting binding occurs only when the last phosphorylation state is reached. The following expression is obtained:

$$CV(\mathrm{E}^*_{\mathrm{tot,area}}|_{t=\infty}) = \frac{\sqrt{\sum_{j=1}^{n}(\nu_j \tau_j)^2}}{\sum_{j=1}^{n} \nu_j \tau_j}. \tag{5.4}$$

In case of mouse, using the biochemistry sequences yielding the results depicted on the left panel of Fig. 8, $CV(\mathrm{E}^*_{\mathrm{tot,area}}|_{t=\infty}) = 0.51$ is obtained. By removing the simplifying assumption previously mentioned, numerical results supplied the slightly higher value $CV(\mathrm{E}^*_{\mathrm{tot,area}}|_{t=\infty}) = 0.55$ [13]. This result has been obtained performing many and many numerical trials in a stochastic setting, such as statistical convergence is reached, and then computing the standard deviation and average of $\mathrm{E}^*_{\mathrm{tot,area}}$. It is here remarked that the CV of $\mathrm{E}^*_{\mathrm{tot}}$ cannot be directly measured through dedicated experiments, but the biochemistry values used for obtaining the results in Fig. 8 have been calibrated through experimental findings and so are fully justified [21]. However, experimentalists are able to determine the CV of $R_{\mathrm{rel,area}}|_{t=\infty}$ which has been estimated equal to 0.36. Thus the rod photoresponse mechanisms operate a variability reduction from the effector to the output current, conferring adequate stability to the SPR thus providing acute night vision. While the variability sources are quite well understood by researchers, the mechanisms operating the variability suppression in the SPR, yielding the low CV experimentally observed for the output current are still debated. One hypothesis was that the variability suppression was operated at the level of the activation cascade, where many and many phosphorylation steps were supposed to exist, yielding a low value of the $\mathrm{E}^*{}_{\mathrm{tot}}$ variability, and identifying this variability with the output current variability. In fact, by keeping constant the total average lifetime of the activated rhodopsin molecule and artificially increasing the number of phosphorylation steps occurring before final quenching, the $CV(\mathrm{E}^*_{\mathrm{tot,area}}|_{t=\infty})$ can be made small enough through formula (5.4). However this hypothesis yields not realistic results, requiring a too high number n of phosphorylation steps. Thus the CV reduction must be operated at the level of the cytosol cascade. In order to investigate about this issue and to evaluate the contribution of each possible candidate in the cytosol cascade to operate the CV reduction, the homogenized model was employed for generating virtually KO rods, each one not expressing certain components of the cytosol cascade [13]. In particular, four different components were investigated:

- the second messengers diffusion (D or d);
- the calcium feedback (F or f);
- the nonlinearities in the flux terms (2.8) on the outer shell (N or n);
- the nonlinear source in (2.6), accounting for the cG hydrolysation operated by activated E^* (S or s),

where the capital letter indicate that the relevant property is enabled whereas the lowercase letter indicate that the relevant property has been removed. In particular, d indicates that the diffusion coefficients are infinite, i.e. that the model becomes

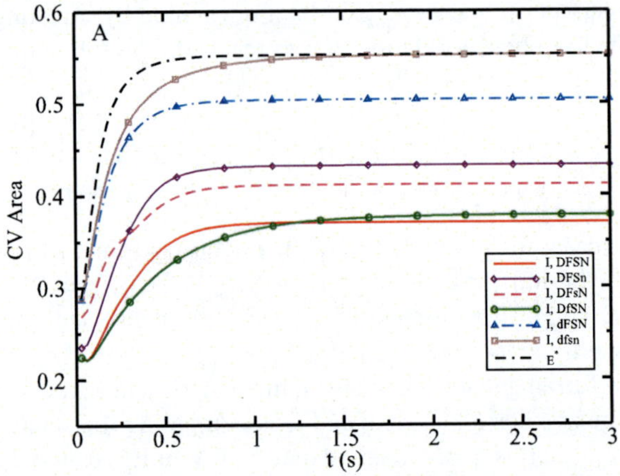

Fig. 18 CV of $R_{\mathrm{rel,area}}$ versus time, relevant to different virtually KO rods. The CV of the corresponding $E^*{}_{\mathrm{tot,area}}$ is also reported for comparison

globally well stirred; *f* indicates that the rod operates under calcium clamped condition, i.e. the Ca^{2+} concentration remains fixed to its dark equilibrium value for all the process; *n* indicates that the flux terms on the outer shell have been linearized; *s* indicates that the cG concentration appearing in the hydrolysis term relevant to E^* is kept fixed to its dark value, thus avoiding any cG local depletion near the activated disc.

The results of the analysis are summarized in Fig. 18, where the CV of $R_{\mathrm{rel,area}}$ are reported, computed over many and many numerical SPR trials, as a function of the time t. The CV of $E^*{}_{\mathrm{rel,area}}$ is also reported for comparison. It is here remarked that each sample is characterized by a certain distribution of E^*, which is used for generating the response of all the various considered KO cases. The results, read for high enough values of t such as to include the whole time course of all the considered trials, reveal that the nonlinearities in the flux terms on the outer shell and in the hydrolysis source term operate the CV reduction, whereas the calcium feedback mechanism contribution is immaterial. The latter occurrence has also an experimental confirmation, since it is possible to measure the SPR in rod mouses under clamped calcium conditions. As a matter of facts, the nonlinear terms reduce the differences in current drop generated by long-living activated rhodopsins and short-living activated rhodopsins, operating the CV reduction from effector E^*_{tot} to rod response R_{rel}. It is observed that if the equations governing the cytosol cascade are linearized, the CV of R_{rel} becomes equal to the CV of E^*_{tot} [13]. Moreover, the smaller the diffusion coefficients, the higher the local cG drop around the activated disc, the higher the CV reduction operated by the nonlinearities in the flux terms, which are a function of cG. In particular, in the cases where the diffusion is set to infinity (*d*) the achieved CV reduction is quite small since the cG drop is spread on all the rod cell, thus resulting very small.

References

1. D. Andreucci, Existence and uniqueness of solutions to a concentrated capacity problem with change of phase. Eur. J. Appl. Math. **1**, 339–351 (1990)
2. D. Andreucci, P. Bisegna, E. DiBenedetto, Homogenization and concentrated capacity in reticular almost disconnected structures. C. R. Acad. Sci. Seri. I Math. **335**, 329–332 (2002)
3. D. Andreucci, P. Bisegna, G. Caruso, H.E. Hamm, E. DiBenedetto, Mathematical model of the spatio-temporal dynamics of second messengers in visual transduction. Biophys. J. **85**, 1358–1376 (2003)
4. D. Andreucci, P. Bisegna, E. DiBenedetto, Homogenization and concentrated capacity for the heat equation with non-linear variational data in reticular almost disconnected structures and applications to visual transduction. Ann. Mat. Pura Appl. **182**(4), 375–407 (2003)
5. D. Andreucci, P. Bisegna, E. DiBenedetto, Mathematical problems in visual transduction phototransduction in vertebrate photoreception, in *Trends in Partial Differential Equations*. Progress in Nonlinear Differential Equations, vol. 81, Special Volume in honor of V. Solonnikov (2005), pp. 65–80
6. D. Andreucci, P. Bisegna, E. DiBenedetto, Homogenization and concentration of capacity in phototransduction in vertebrate photoreception. Appl. Anal. **85**(1–3), 303–331 (2006), Special Issue in honor of C. Pucci
7. D. Andreucci, P. Bisegna, E. DiBenedetto, Homogenization and concentration of capacity in rod outer segments with incisures. Appl. Anal. **85**(1–3), 303–331 (2006)
8. A. Bensoussan, J.L. Lions, G. Papanicolau, *Asymptotic Analysis for Periodic Structures* (North-Holland, New York, 1978)
9. P. Bisegna, G. Caruso, D. Andreucci, L. Shen, V.V. Gurevich, H.E. Hamm, E. DiBenedetto, Diffusion of the second messengers in the cytoplasm acts as a variability suppressor of the single photon response in vertebrate phototransduction. Biophys. J. **94**, 3363–3383 (2008)
10. G. Caruso, H. Khanal, V. Alexiades, F. Rieke, H.E. Hamm, E. DiBenedetto, Mathematical and computational modeling of spatio-temporal signaling in rod phototransduction. IEE Proc. Syst. Biol. **152**, 119–137 (2005)
11. G. Caruso, P. Bisegna, L. Shen, D. Andreucci, H.E. Hamm, E. DiBenedetto, Modeling the role of incisures in vertebrate phototransduction. Biophys. J. **91**, 1192–1212 (2006)
12. G. Caruso, P. Bisegna, L. Lenoci, D. Andreucci, V.V. Gurevich, H.E. Hamm, E. DiBenedetto, Kinetics of rhodopsin deactivation and its role in regulating recovery and reproducibility in wild type and transgenic mouse photoresponse. PLoS Comput. Biol. (2010)
13. G. Caruso, P. Bisegna, L. Lenoci, D. Andreucci, V.V. Gurevich, H.E. Hamm, E. DiBenedetto, Identification of key factors that reduce the variability of the single photon response. PNAS **108**(19), 7804–7807 (2011)
14. Ph.G. Ciarlet, V. Lods, Asymptotic analysis of linearly elastic shells. I. Justification of membrane shell equations. Arch. Ration. Mech. Anal. **136**, 119–161 (1996)
15. D. Cioranescu, J. Saint Jean Paulin, Homogenization in open sets with holes. J. Math. Anal. Appl. **71**, 590–607 (1979)
16. M. Gray-Keller, W. Denk, B. Shraiman, P.B. Detwiler, Longitudinal spread of second messenger signals in isolated rod outer segments of lizards. J. Physiol. **519**, 679–692 (1999)
17. O.P. Gross, E.N. Pugh Jr, M.E. Burns, Spatiotemporal cGMP dynamics in living mouse rods. Biophys. J. **102** (2012), 1775–1784.
18. T.D. Lamb, E.N. Pugh Jr, A quantitative account of the activation steps involved in phototransduction in amphibian photoreceptors. J. Physiol. **449**, 719–758 (1992)
19. E. Magenes, Stefan problems with a concentrated capacity. Boll. Unione Mat. Ital. Sez. B Artic. Ric. Mat. **1**(1), 71–81 (1998)
20. O.A. Oleinik, A.S. Shamaev, G.A. Yosifian, *Mathematical Problems in Elasticity and Homogenization*. Studies in Mathematics and its Applications, vol. 26 (North Holland, Amsterdam, 1992)

21. L. Shen, G. Caruso, P. Bisegna, D. Andreucci, V.V. Gurevich, H.E. Hamm, E. DiBenedetto, Dynamics of mouse rod phototransduction and its sensitivity to variation of key parameters. IET Syst. Biol. **4**(1), 12–32 (2008)
22. O.C. Zienkiewicz, R.L. Taylor, J.Z. Zhu, *The Finite Element Method: Its Basis and Fundamentals* (Butterworth-Heinemann, Amsterdam, 2005)

Existence, Decay Time and Light Yield for a Reaction-Diffusion-Drift Equation in the Continuum Physics of Scintillators

Fabrizio Daví

Abstract A scintillator is a material which converts incoming ionizing energy into visible light. This conversion process, which is a strongly nonlinear one, can be described by a Reaction-Diffusion-Drift equation we obtain from a model of continua with microstructure endowed with a suitable thermodynamics. For such an equation it can be show the global existence of renormalizable and weak solutions, and the solutions exponential decay estimates can be given; moreover we give also a mathematical definition for the light yield which is a measure of scintillation efficiency.

Keywords Reaction-Diffusion-Drift models · Scintillators · Entropy methods · Exponential rate of convergence

1 Introduction

A scintillator crystal is a "wavelength shifter" which converts energy, typically γ-rays, into photons in the frequency range of visible light. For this reason scintillators crystals are used in high-energy physics and in medical and security applications [1]. The physics of scintillation is quite complex but it can be conveniently divided in three major phenomena which correspond to three different time and space scales: (1) the incoming energy generates an ionized region of few nanometers populated by charged energy carriers, a scale which we call the *Microscopic* dealing with the creation of excitation carriers in the ionized region; (2) these energy carriers generates other energy carriers within a greater region: when these carriers recombine a part of them generate photons. Such a phenomena evolves at a *Mesoscopic* scale. (3) the light rays propagate within the crystal, a phenomena which happens at a *Macroscopic* scale.

F. Daví (✉)
DICEA, Universitá Politecnica delle Marche, Ancona, Italy
e-mail: davi@univpm.it

V. Vespri et al. (eds.), *Harnack Inequalities and Nonlinear Operators*,
Springer INdAM Series 46, https://doi.org/10.1007/978-3-030-73778-8_6

In [2] we obtained, by means of a continuum with microstructure theory and a suitable thermodynamics, a model which describes the phenomena at the mesoscopic scale in terms of a Reaction-Diffusion Drift equation for the energy carriers descriptors, with Neumann-type boundary conditions, coupled with the heat and electrostatic equations.

Such equation and its associated variational formulation are the starting point for the correct mathematical description of the two most important physical parameters which characterize a scintillator crystal: the *Light Yield* Y, which is the ratio between the collected light energy and the energy of the incoming ionizing radiation (and which is indeed a measure of the scintillator efficiency) and the *Scintillation Decay time* τ which is the time required for scintillation emission to decrease to e^{-1} of its maximum and is a measure of the scintillator resolution.

Here first we show the main results obtained into [2] and then, by adapting the results of [3] to the present formulation, we proof global existence of renormalized and weak solutions. Then by following the approach and the ideas of [4] (*vid. also* [5]) we give an estimate for the decay time. Finally, we propose a suitable definition for the light yield based on the decay time estimate.

2 The Evolution Equation for Scintillators

2.1 The Excitation Carrier Density Vector

In order to define the basic state variable for our problem, we have to deal briefly with the scintillation phenomena at the microscopic scale and to the features of it which appears at the mesoscopic scale: we give here only the main ideas. More details can be found in [2]. Basically, the incoming energy E^* which hits the crystal at a point x generates a great number of excitation carriers within a cylindrical track of radius r and energy-dependent length $L = L(E^*)$ from x: on this track we define an excitation density [6].

In [2] we show that the relevant descriptor of the microscopic phenomena is the excitation density times the area of the cross-section of the cylindrical track and then, by "zooming-out" to the mesoscopic volume Ω centered on x we get, by means of renormalization techniques, the mesoscopic descriptor in Ω:

$$N = \frac{1}{\pi r^2 L(E^*)} \frac{E^*}{E_{exc}} > 0\,, \tag{1}$$

where E_{exc} is the *excitation energy*, which depends on the specific scintillator crystal.

Since the excitation carriers may exhibits different physical behaviour, *i.e.* can recombine into photons or other kind of excitation carriers rather then annihilate themselves in different ways, then N can be decomposed into the sum of k different kind of excitation carriers, the value of k depending on how much we want a more detailed description of phenomena (for instance $k = 2$ in [7] whereas $k \geq 11$ in [8]):

$$N = \sum_{i=1}^{k} n_i\,, \quad n_i > 0\,. \tag{2}$$

We find useful to introduce an *excitation carrier density* k-dimensional vector as the basic state variable for our theory:

$$n \equiv (n_1\,, n_2\,, \ldots\,, n_k)\,, \quad (x\,, t) \mapsto n_j(x\,, t) > 0\,, \quad j = 1\,, \ldots k\,; \tag{3}$$

where $(x\,, t) \in \Omega \times [0\,, \tau)$ with Ω a mesoscopic control volume and $n \in \mathcal{M} \equiv (0\,, \infty)^k$.

Let q^* be the charge density associated to the incoming energy E^*, then such a charge and the excitation carriers generate an *electric potential* $(x\,, t) \mapsto \varphi(x\,, t)$:

$$-\,\epsilon_o \Delta\varphi = q^* + e\,z \cdot n\,, \quad \text{in}\,\Omega \times [0\,, \tau)\,, \tag{4}$$

with Neumann-type boundary conditions on $\partial\Omega \times [0\,, \tau)$: here e is the elementary electron charge, $\epsilon_o > 0$ is the material permittivity and $z \in \mathbb{Z}^k$ is the *charge vector*, which accounts for the sign and the amount of electric charge e of each carrier.

2.2 The Reaction-Diffusion-Drift Equation

To model the recombination of excitation carriers within Ω, in [2] we wrote the equation of electric current balance in terms of the theory of continua with microstructure (*vid. e.g.* [9]) endowed with a suitable thermodynamic and appropriate constitutive hypotheses. In particular we assumed a Gibbs free-energy

$$\psi = \varepsilon - \theta\eta \tag{5}$$

with internal energy ε and entropy η given respectively by:

$$\varepsilon(n\,, \theta) = e\varphi\,z \cdot n + u(\theta)\,, \tag{6}$$

and

$$\eta(n\,, \theta) = -k_B \sum_{i=1}^{k} n_i(\log C_i n_i - 1) + \lambda \log\theta\,, \tag{7}$$

where k_B is the Boltzmann constant, $\lambda > 0$ the latent heat and $C_i > 0$ are normalizing constant. In the model we obtained, the only interaction with the macroscopic scale is the *absolute temperature* $\theta = \theta(x, t) > 0$: a wider range of macroscopic variables, like mechanical strain, crystal defects and electromagnetic fields (as in [10], *e.g.*) will be dealt with in future papers.

From the dissipation inequality, constitutive assumptions and the balance laws for a continuum with microstructure we arrive at an equation which describes the generation and recombination of excitation carriers:

$$\operatorname{div}(D\nabla n + MNz \otimes \nabla\varphi) - K(n)n = \dot{n}\,, \quad \text{in } \Omega \times [0, \tau)\,, \tag{8}$$

which is a reaction-diffusion-drift equation with Neumann boundary condition on $\partial\Omega \times [0, \tau)$ and initial conditions $n(x, 0) = n_0(x)$ which depend, by means of (1) and (2), on the incoming energy E^* at x.

The various terms in (8) represent:

- N is the $k \times k$ matrix $N = \operatorname{diag}(n_1, n_2, \dots, n_k)$;
- $M = M(\theta)$ is the $k \times k$ symmetric and positive-definite *Mobility matrix*;
- $D = (k_B\theta/e)M$, is the $k \times k$ *Diffusivity matrix*;
- φ is the *local electric field* solution of (4);
- $K = K(n, \theta)$ is a non-linear function of n which describes the recombination process.

Equation (8) is coupled with (4) and with the heat equation (with an electrostatic source term)

$$\dot{\theta} = \operatorname{div} \mathbf{C}\nabla\theta - ez \cdot \dot{n}\,, \quad \text{in } \Omega \times [0, \tau)\,, \tag{9}$$

with Neumann boundary conditions on $\partial\Omega \times [0, \tau)$; here $\mathbf{C}$ is the positive-definite crystal *Conductivity tensor*.

Equation (8) generalizes the two most important phenomenological models for scintillation, namely the *Kinetic* (*vid. e.g.* [11]) and the *Diffusion* models [12]: they are the same equations postulated into [8] and used into [13] to perform numerical analysis of solutions; they are also identical (apart for the reaction term $K(n)n$) to the equations for the semiconductors obtained using a different approach, into [14, 15] and [16].

Thermodynamics allows to write the *Dissipation* associated to Eq. (8):

$$\mathcal{D} = 2\Psi(n, \mu, \theta)\,, \tag{10}$$

where the *Conjugate dissipation functional* is defined as

$$\Psi(n, \mu, \theta) = \frac{1}{2}\int_\Omega S(n, \theta)[\nabla\mu] \cdot \nabla\mu + H(n, \theta)\mu \cdot \mu > 0\,; \tag{11}$$

here μ is the *Scintillation potential* (indeed the equivalent for scintillators of the electrochemical potential in semiconductors)

$$\mu = \frac{\partial \psi}{\partial n} = e\varphi z + k_B\theta \log(n^*) , \quad \log(n^*) \equiv (\log C_1 n_1 , \dots \log C_k n_k) , \tag{12}$$

and where the positive-definite $k \times k$ matrices S and H are given by $S(n , \theta) = e^{-1}M(\theta)N(n)$ and $H(n , \theta)\mu = K(n , \theta)n$.

We notice that relation (12) can be inverted to obtain

$$n = \Lambda\, c , \tag{13}$$

with $\Lambda = (e^{\frac{1}{k_B\theta}(\mu_1 - ez_1\varphi)}, \dots , e^{\frac{1}{k_B\theta}(\mu_k - ez_k\varphi)})$ and $c \equiv (c_1 , \dots , c_k)$ with $c_i = C_i^{-1}$.

By means of (11), Eq. (8) can be put in the equivalent gradient flow formulation, namely:

$$\dot{n} = -D\Psi(n , \mu , \theta) , \tag{14}$$

where $D\Psi$ denotes the Frechet derivative of the dissipation Ψ; notice that (8) can be expressed in terms of the scintillation potential as:

$$\text{div}\ S[\nabla\mu] - H\mu = \dot{n} , \tag{15}$$

a form we shall make use of in the sequel.

3 Existence, Decay Time Estimates and Light Yield

Trought this section we shall deal with isothermal scintillators, in such a way that the fixed temperature $\theta = \theta_o$ appears only as a parameter in the constitutive terms M and K and the problem is described by Eqs. (4) and (8) only: moreover we shall assume that the domain can be rescaled by a characteristic length l^* and w.l.o.g. we set that the adimensional parameter

$$\beta = \frac{k_B\theta_o}{el^*} = 1 , \tag{16}$$

in such a way that $D = M$. With a slight abuse of notation we shall still denote Ω the rescaled domain.

3.1 Global Existence

The problem of finding existence, asymptotic estimates and qualitative bounds for the solutions for the reaction-diffusion drift equations like (8) coupled with (4) has received a strong attention in the recent years, *vid. e.g.* [3–5, 17–29]: to this regard it is important to remark that most of them deal with semiconductors or chemical reactions which differ from scintillators by the reaction term $K(n)n$.

In [3] however, a global existence for (8) with Neumann-type boundary conditions in terms of *renormalized solutions* was obtained for a general reaction term $K(n)n$. According to the definition given into [29], renormalized solutions n to the reaction-diffusion-drift equation (8) are defined by the condition that for all functions $\xi : \mathcal{M} \to \mathbb{R}$ with compactly supported derivative $\nabla_n \xi$, the function $\xi(n)$ must satisfy the equation derived from (8) by a formal application of the chain rule in a weak sense.

More precisely, according to [3] and [29], $n \equiv (n_1, n_2, \ldots, n_k)$ is a *renormalized solutions* for (8) if $\forall \tau > 0$, $n_i \in L^2([0, \tau); H^1(\Omega))$ and for any $\xi \in C^\infty(\mathcal{M})$ satisfying $\nabla_n \xi \in C_0^\infty(\mathcal{M}; \mathbb{R}^k)$ and $\psi \in C_0^\infty(\bar{\Omega} \times [0, \tau))$ it holds:

$$
\begin{aligned}
&\int_0^\tau \int_\Omega \xi(n)\dot{\psi} + \int_\Omega \xi(n)\psi\big|_0^\tau = \\
&+ \int_0^\tau \int_\Omega (\nabla\nabla\xi \cdot S[\nabla\mu] \otimes \nabla n)\psi + \int_0^\tau \int_\Omega S[\nabla\mu]\nabla\xi \cdot \nabla\psi \\
&+ \int_0^\tau \int_\Omega (H\mu \cdot \nabla\xi)\psi, .
\end{aligned}
\tag{17}
$$

Let $\mathcal{E}$ be the *total scintillation entropy* on the control volume Ω:

$$
\mathcal{E}(n) = -\int_\Omega \sum_{i=1}^k n_i(\log C_i n_i - 1)\,, \tag{18}
$$

then the main result of [3] rephrased in terms of (8) states that, provided the following hypotheses hold:

(H1) Drift term: $\nabla\varphi \in L^\infty([0, \tau); L^\infty(\Omega))$;
(H2) Reaction term: $K(n)n \in C_0([0, \tau)^k; \mathcal{M})$;
(H3) Initial data: $n_o \equiv (n_1^0, n_2^0, \ldots, n_k^0)$ is measurable, $n_i^0 > 0$ in Ω, $i = 1, 2, \ldots k$ and

$$
\mathcal{E}(n_o) < +\infty\,;
$$

(H4) There exist numbers $\pi_i > 0$ and $\lambda_i \in \mathbb{R}$, $i = 1, 2, \dots, k$ such that for all $n \equiv (n_1, n_2, \dots, n_k) \in \mathcal{M}$, the following inequality holds:

$$\sum_{i=1}^{k} \pi_i (K(n)n)_i (C_i \log n_i + \lambda_i) \leq 0\,;$$

(H5) The mobility matrix M is symmetric and positive-definite;

then Eq. (8) admits a renormalized solution $n \equiv (n_1, n_2, \dots, n_k)$ satisfying $n_i > 0$ in Ω, $i = 1, 2, \dots k$ and

$$\mathcal{E}(n) =< +\infty\,; \qquad \forall t > 0\,. \tag{19}$$

As pointed out in [29], moreover, any renormalized solution for which $K(n)n = H\mu \in L^1([0\,,\infty)^k; \mathcal{M})$ is also a weak solution of (14) in the sense that, for any $v \equiv (v_1, \dots v_k) \in C^\infty([0\,,\tau)^k\,, \mathcal{M})$:

$$\int_\Omega (v \cdot n)|_0^\tau - \int_0^\tau \int_\Omega n \cdot \dot{v} = - \int_0^\tau \int_\Omega S[\nabla\mu] \cdot \nabla v + H\mu \cdot v\,; \tag{20}$$

as far as we know we can instead say nothing about the global existence in time of smooth solutions.

3.2 Decay Time

The available experimental data (*vid. e.g.* the recent analysis in [30]) and the numerical solution of phenomenological models as in [13], show that the excitation carriers decay exponentially to an asymptotic value n_∞, namely:

$$\|n(\cdot\,, t) - n_\infty(\cdot)\| = A_f \exp(-t/\tau_f) + A_s \exp(-t/\tau_s)\,, \tag{21}$$

where the indeces f and s denotes the so-called *fast* and *slow* components of the excitation, respectively. Accordingly, since by definition the *Decay time* is the time required for scintillation emission to decrease to e^{-1} of its maximum, then we get a Fast Decay Time τ_f and a Slow Decay Time τ_s.

In many cases one of the components is negligible and the decay obeys a simple exponential law, which can be also used to describe an average decay time. To this regard, in [4, 5, 20, 27] and [28], an explicit estimate of the asymptotic convergence was obtained for the cases of chemical reactions and semiconductors. In particular in [4] the Rosbroeck model with Shockley-Read-Hall potential for semiconductors was studied. In the following we shall see how the same approach and ideas of [4] can be extended to the case of scintillators in order to obtain an explicit estimate for the decay time.

Scintillation depends on the evolution of charge carriers: accordingly we must require that through the whole process the electric charge is conserved. Accordingly, let $\mathcal{Q} \in \mathbb{R}$ be the *total electric charge*

$$\mathcal{Q}(n) = \mathcal{Q}^* + \int_\Omega ez \cdot n \,, \quad \mathcal{Q}^* = \int_\Omega q^* \,, \tag{22}$$

then by (4) with Neumann boundary condition we must have:

$$\mathcal{Q}^* + \int_\Omega ez \cdot n = 0 \,, \quad \forall t \in [0 \,, \tau) \,. \tag{23}$$

We remark that (23) is the necessary condition to have an unique weak solution $\varphi \in H^1(\Omega)$ to Eq. (4) with Neumann boundary conditions and such that $\overline{\varphi} = 0$, where

$$\overline{f} = \frac{1}{\mathrm{vol}\,\Omega} \int_\Omega f \,,$$

denotes the mean value on Ω.

Moreover (23) and the charge conservation lead to

$$\frac{\mathrm{d}}{\mathrm{d}t}\mathcal{Q} = \int_\Omega ez \cdot \dot{n} = 0 \,, \quad \forall t \in [0 \,, \tau) \,, \tag{24}$$

and from $(24)_1$ and (8) with Neumann-type boundary conditions:

$$\int_\Omega K(n)n \cdot z = 0 \,, \quad \forall t \in [0 \,, \tau) \,. \tag{25}$$

It is important to remark that the total charge $\mathcal{Q}$ depends on the type of ionizing radiation which hits the scintillator: indeed for γ- and X-rays we have $\mathcal{Q}^* = 0$, whereas for α-rays it is $\mathcal{Q}^* > 0$ and $\mathcal{Q}^* < 0$ for β-rays.

Let $n_\infty(x)$ and $\varphi_\infty(x)$ be the stationary solution(s) of (8) and (4), i.e. with $\dot{n} = 0$ (cf. [21]); it is easy to see from (15) that for the stationary solutions the scintillation potential vanishes i.e.

$$\mu_\infty = 0 \,. \tag{26}$$

Accordingly, from (13), by (12) we have:

$$n_\infty = F_\infty c \,, \tag{27}$$

where $F_\infty = \text{diag}(e^{-ez_1^*\varphi_\infty}, \dots, e^{-ez_k^*\varphi_\infty})$, $z_j^* = ez_j/k_B\theta_o$ and with φ_∞ the unique solution of the Neumann-type problem [21]

$$-\,\epsilon_o\Delta\varphi_\infty = q^* + e\,z\cdot F_\infty c\,,\quad \text{in}\,\Omega\,,\quad \overline{\varphi}_\infty = 0\,, \tag{28}$$

provided

$$-\,\mathcal{Q}^* = ez\cdot\int_\Omega n_\infty = ez\cdot\overline{F}_\infty c\,, \tag{29}$$

holds; we remark that condition (26) trivially verifies both (25) and:

$$z\cdot\overline{K(n_\infty)n_\infty} = 0\,. \tag{30}$$

We notice that, in the case of γ-rays we have $\mathcal{Q}^* = 0$ and condition (29) implies that, for $\mathcal{S} \equiv \text{span}\{(\overline{F}_\infty)^T z\}$, then $c \in \mathcal{S}^\perp$.

In order to grant uniqueness for c and hence for $(n_\infty\,,\varphi_\infty)$, we need additional hypotheses on the reaction term $K(n)n$, as it was done in [4] for the case of Rosbroeck semiconductors with $k = 2$, where for the reaction term was assumed a Shockley-Read-Hall potential (*vid. also* [28]). Here we simply assume as a constitutive prescription that the reaction term $K(n)$ is such that c is unique and there exists a constants K_1 such that:

$$0 < K_1 \le \|K(n)\|_{L^\infty(\Omega)}\,; \tag{31}$$

then we may assume that the following bounds for c and n_∞ hold (cf.[4]):

$$\|c\| \le e^{\Phi_\infty}(1+|\mathcal{Q}^*|)\,,\qquad \|n_\infty\| \le e^{2\Phi_\infty}(1+|\mathcal{Q}^*|)\,, \tag{32}$$

with $\Phi_\infty = e\|z\varphi_\infty\|_{L^\infty(\Omega)}$.

The total Gibbs free-energy for a scintillator is given by

$$\mathcal{G}(n\,,\varphi(n)) = \int_\Omega \psi(n\,,\varphi(n)) = \int_\Omega \varepsilon(n\,,\varphi(n)) - \theta_o\eta(n\,,\varphi(n))\,; \tag{33}$$

then following [4] (*see also* [29]), we define the *Relative Gibbs free-energy* (which in [4] is referred as "relative entropy") as:

$$\mathcal{G}(u|v) = \mathcal{G}(u) - \mathcal{G}(v) - D\mathcal{G}(v)(u-v)\,. \tag{34}$$

By an explicit calculation we obtain:

$$\mathcal{G}(n|n_\infty) = \int_\Omega \sum_{i=1}^k n_i\log(\frac{n_i}{n_i^\infty}) + (n_i^\infty - n_i) + \frac{1}{2}\varepsilon_o\|\nabla\varphi - \nabla\varphi_\infty\|^2\,, \tag{35}$$

and then, by an easy calculation, it can be shown that the *Dissipation* $\mathcal{D}$ is given by [2]:

$$\mathcal{D} = -\frac{\mathrm{d}}{\mathrm{d}t}\mathcal{G} = 2\Psi(n\,,\mu) > 0\,. \tag{36}$$

We follow [4] and by starting from (36), by means of a repeated use of Csiszár-Kullback-Pinsker type inequalities we may arrive, provided (32) hold, to the following estimates for the case $k = 2$:

$$\begin{aligned} &\mathcal{D}(n\,,\varphi(n)) \geq C_1\mathcal{G}(n\,,\varphi(n))\,,\\ &\|n - n_\infty\|^2_{L^1(\Omega)} + \|\varphi - \varphi_\infty\|^2_{H^1(\Omega)} \leq C_2\mathcal{G}(n_o\,,\varphi_o)e^{-C_1 t}\,, \end{aligned} \tag{37}$$

with φ_o the unique solution of

$$-\,\epsilon_o\Delta\varphi_o = q^* + e\,z\cdot n_o\,, \quad \text{in}\,\Omega\,, \quad \overline{\varphi}_o = 0\,, \tag{38}$$

with Neumann-type boundary conditions and where the parameters $C_{1,2}$ have the explicit expression [4]:

$$\begin{aligned} C_1^{-1} &= \frac{1}{2}e^{2\Phi_\infty}(1 + |\mathcal{Q}^*|)\max\{\frac{\epsilon}{M^*}e^{2\Phi_\infty}(1 + |\mathcal{Q}^*|)\,, \frac{1}{K_1}\}\,\cdot\\ &\cdot(1 + \frac{L(\Omega)}{\epsilon}e^{2\Phi_\infty}(1 + |\mathcal{Q}^*|))\,,\\ C_2 &= (3e^{2\Phi_\infty}(1 + |\mathcal{Q}^*|) + \frac{1}{2}\mathcal{G}(n_o\,,\varphi_o) + \frac{2}{\epsilon}(1 + L(\Omega))\,, \end{aligned} \tag{39}$$

where $L(\Omega)$ is the Poincaré constant of Ω, M^* is the smallest eigenvalue of M and $\mathcal{Q}^*$, ϵ are normalized with respect to the elementary charge.

The expression for the decay time $\tau = C_1^{-1}$ depends, by $(39)_1$, in an explicit manner on the mobility M, the reaction term $K(n)$, the initial data n_o, the charge $\mathcal{Q}^*$ and the scintillation volume Ω. The extension to the case $k > 2$ and to specific expression for $K(n)$ will be the object of further studies: however, as far as we know, this is the first explicit estimate of the decay time in term of the problem physical (and measurable) parameters.

3.3 *Light Yield*

In order to define the light yield we must be able to discriminate the recombinations of excitation carriers which converts into photons from those which exhibit "quenching", that is recombination without emission. To this regard in the most successful phenomenological model for scintillator, the "Kinetic model", borrowed from chemical reactions (*vid. e.g.* [1, 8, 11]), the matrix $K(n)$ was assumed as a quadratic function of n:

$$K_{ij}(n) = R_{ij} + G_{ij} + E_{ij} + (\mathsf{R}_{ijh} + \mathsf{G}_{ijh})n_h + \mathbb{G}_{ijhm}n_h n_m \,, \tag{40}$$

$i, j, h, m = 1, 2, \ldots k$, where the terms R_{ij} and R_{ijh} account for the linear and quadratic recombination, the terms G_{ij}, G_{ijh} and $\mathbb{G}_{ijhm}$ accounts for the linear, quadratic and cubic (Auger) quenching whereas the exchange matrix E_{ij} accounts for the excitation carriers which converts in other types. We remark that in this case bound (31) hold for $m = 2$ with $K_1 \approx \|R + G + E\|$ and $K_2 \approx \|\mathsf{R} + \mathsf{G}\| + \|\mathbb{G}\|$.

The most accepted definition of light yield in terms of the parameters of the phenomenological model is given e.g. in [31]; let $n_p(x\,, t)$ be the solution of (8) for $G_{ij} = 0$, $\mathsf{G}_{ijh} = 0$ and $\mathbb{G}_{ijhm} = 0$, i.e. the solution which converts into visible light photons and let:

$$N_p(x\,, t) = \sum_{j=1}^{k} n_j^p(x\,, t)\,, \quad N_o(x) = \sum_{j=1}^{k} n_j^o(x)\,, \tag{41}$$

(here $n_j^o(x)$ are the component of the initial data $n_o(x)$); then we define the *Local light yield* Y_L at a given point $\bar{x}$ as:

$$Y_L(\bar{x}) = \frac{1}{\bar{\tau} N_o(\bar{x})} \int_0^{\bar{\tau}} N_p(\bar{x}\,, t)\mathrm{d}t\,, \tag{42}$$

where the characteristic time $\bar{\tau}$ is sometimes assumed as

$$\bar{\tau}^{-1} = \sup\{R_{ij}\}\,. \tag{43}$$

A *Global light yield* can be defined taking into account a characteristic volume, either the volume of the track (as in [11]) or the scintillation volume Ω about $\bar{x}$:

$$Y(\Omega) = \int_\Omega \frac{1}{\bar{\tau} N_o(\bar{x})} \int_0^{\bar{\tau}} N_p(\bar{x}\,, t)\mathrm{d}t\,. \tag{44}$$

We propose here a different definition for the global light yield, based on the results of the previous section. Let

$$\bar{N}_o = \|n_o(x) - n_\infty(x)\|_{L^1(\Omega)}\,, \quad \bar{N}_p(t) = \|n_p(x\,, t) - n_\infty(x)\|_{L^1(\Omega)}\,, \tag{45}$$

then the bound (37) holds and we may define an estimate for the global light yield:

$$Y(\Omega) = \frac{1}{\bar{\tau}\bar{N}_o} \int_0^{\bar{\tau}} \bar{N}_p(t)\mathrm{d}t \le \frac{1}{\bar{\tau}\bar{N}_o} \int_0^{\bar{\tau}} C_2 \mathcal{G}(n_o\,, \varphi_o) e^{-C_1 t}\mathrm{d}t\,, \tag{46}$$

to arrive at:

$$Y(\Omega) \leq \frac{1}{\bar{\tau}\bar{N}_o}\frac{C_2}{C_1}(1 - e^{-C_1\bar{\tau}})\mathcal{G}(n_o\,, \varphi_o)\,, \tag{47}$$

with $C_{1,2}$ evaluated for $G_{ij} = 0$, $\mathsf{G}_{ijh} = 0$ and $\mathbb{G}_{ijhm} = 0$. A further analysis of such definition and the study of its relation with the classical one will be done in the future.

Acknowledgments The research leading to these results is within the scope of CERN R&D Experiment 18 "Crystal Clear Collaboration" and has received funding from the European Research Council under the COST action TD-1401 "FAST—Fast Advanced Scintillation Timing". The author wishes to thanks K. Fellner for pointing his attention on Refs. [4] and [28].

References

1. P. Lecoq, A. Gektin, M. Korzhik, *Inorganic Scintillators for Detector Systems: Physical Principles and Crystal Engineering*, 2nd ed. (Springer, Berlin-Heidelberg, 2017)
2. F. Daví, A continuum theory for scintillating crystals. Eur. J. Phys. B **92**, 1–16 (2019). https://doi.org/10.1140/epjb/e2018-80590-2
3. X. Chen, A. Jüngel, Weak-strong uniqueness of renormalised solutions to reaction-cross diffusion systems. Math. Models Methods Appl. Sci. **29**(2), 237–270 (2019)
4. K. Fellner, M. Kniely, On the entropy method and exponential convergence to equilibrium for a recombination-drift-diffusion system with self-consistent potential. Appl. Math. Lett. **79**, 196–204 (2018)
5. L. Desvillettes, K. Fellner, Exponential decay toward equilibrium via entropy methods for reaction-diffusion equations. J. Math. Anal. Appl. **319**, 157–176 (2006)
6. J.E. Jaffe, Energy and length scales in scintillator nonproportionality. Nuclear Instrum. Methods Phys. Res. **A570**, 72–83 (2007)
7. Q. Li, J.Q. Grim, R.T. Williams, G.A. Bizarri, W.W. Moses, A transport-based model of material trends in nonproportionality of scintillators. J. Appl. Phys. **109**, 123716 (2011)
8. A. Vasil'ev, From luminescence non-linearity to scintillation non-proportionality. IEEE Trans. Nuclear Sci. **55**(3), 1054–1061 (2008)
9. G. Capriz, *Continua with microstructure*. Springer Tracts in Natural Philosophy (Springer, Berlin, 1989)
10. Y. Xiao, K. Bhattacharya, A continuum theory of deformable, semiconducting ferroelectrics. Arch. Rat. Mech. Anal. **189**, 59–95 (2008)
11. G. Bizzarri, W.W. Moses, J. Singh, A.N. Vasil'ev, R.T. Williams, An analytical model of nonproportional scintillator light yield in terms of recombination rates. J. Appl. Phys. **105**, 044507 (2009)
12. W. Moses, G. Bizzarri, R.T. Williams, S.A. Payne, A.N. Vasil'ev, J. Singh, Q. Li, J.Q. Grim, W.S. Chong, The origins of scintillator non-proportionality. IEEE Trans. Nucl. Sci. **59**(5), 2038–2044 (2012)
13. X. Lu, S. Gridin, R.T. Williams, M.R. Mayhugh, A. Gektin, A. Syntfeld-Kazuch, L. Swiderski, M. Moszynski, *Energy-dependent scintillation pulse shape and proportionality of decay components for CsI:Tl: modeling with transport and rate equations*. Phys. Rev. Applied **7**, 1 (2017). 014007–014030
14. G. Albinus, H. Gajewski, R. Hünlich, Thermodynamic design of energy models of semiconductor devices. Nonlinearity **15**, 367–383 (2002)

15. A. Mielke, A gradient structure for reaction-diffusion systems and for energy-drift diffusion systems. Nonlinearity **24**, 1329–1346 (2011)
16. A. Mielke, On thermodynamical coupling of quantum mechanics and microscopic systems, in Proceedings of the QMath12 Conference (2015), pp. 331–347
17. H. Gajewski, On the uniqueness of solutions to the drift-diffusion model of semiconductor devices. Math. Methods Appl. Sci. **4**(1), 121–133 (1994)
18. A. Glitzky, R. Hünlich, Energetic estimates and asymptotics for electro-reaction-diffusion systems. ZAMM (Z. Angew. Math. Mech.) **77**(11), 823–832 (1997)
19. A. Glitzky, R. Hünlich, Global estimates and asymptotics for electro reaction diffusion systems in heterostructures. Appl. Anal **66**(3–4), 205–226 (1997)
20. L. Desvillettes, K. Fellner, Entropy methods for reaction-diffusion equations: slowly growing a-priori bounds. Rev. Mat. Iberoamericana **24**(2), 407–431 (2008)
21. H. Wu, P. Markowich, S. Zheng, Global existence and saymptotic behavior for a semiconductor drift-diffusion-Poisson model. Math. Methods Appl. Sci. **18**(3), 443–487 (2008)
22. A. Glitzky, Exponential decay of the free energy for discretized electro-reaction-diffusion systems. Nonlinearity **21**, 1989–2009 (2008)
23. M. Bisi, L. Desvillettes, G. Spiga, Exponential convergence to equilibrium via Lyapounov functionals for reaction-diffusion equations arising from non-reversible chemical kinetics. ESAIM: M2AN **43**(1), 151–172 (2009)
24. A. Mielke, R. Rossi, G. Savaré, Variational convergence of gradient flows and rate-independent evolutions in metric spaces. Milan J. Math. Vol. **80**, 381–410 (2012)
25. L. Desvillettes, K. Fellner, *Duality- and entropy methods for reversible reaction-diffusion equations with degenerate diffusion*. Math. Methods Appl. Sci. **38**(16), 3432–3443 (2015)
26. J. Haskovec, S. Hittmeir, P. Markowich, A. Mielke, Decay to equilibrium for energy-reaction-diffusion systems (2016). Preprint. ArXiv1602.05696, 1–40
27. K. Fellner, B.Q. Tang, Convergence to equilibrium of renormalised solutions to nonlinear chemical reaction-diffusion systems (2017). Preprint. ArXiv1708.01427, 1–25
28. K. Fellner, M. Kniely, Uniform convergence to equilibrium for a family of drift-diffusion models with trap-assisted recombination and the limiting Shockley-Read-Hall model. J. Elliptic Parabolic Equ. **6**(2), 529–598 (2020)
29. J. Fischer, Weak-strong uniqueness of solutions to entropy-dissipating reaction-diffusion equations. Nonlinear Anal. **159**(2017), 181–207. https://doi.org/10.1016/j.na.2017.03.00
30. L. Swiderski, M. Moszynski, A. Syntfeld-Kazuch, M. Szawlowski, T. Szczesniak, Measuring the scintillation decay time for different energy depositions in NaI:Tl,LSO:Ce and $CeBr_3$ *scintillators*. Nuclear Instrum. Methods Phys. Res. **A749**, 68–73 (2014)
31. G. Bizzarri, W.W. Moses, J. Singh, A.N. Vasil'ev, R.T. Williams, The role of different linear and non-linear channels of relaxation in scintillator non-proportionality. J. Luminescence **129**, 1790–1793 (2009)

Boundary Harnack Type Inequality and Regularity for Quasilinear Degenerate Elliptic Equations

Giuseppe Di Fazio, Maria Stella Fanciullo, and Pietro Zamboni

Abstract We prove Harnack inequality and global regularity results for weak solutions of a quasilinear degenerate equation in divergence form under natural growth conditions. The degeneracy is given by a suitable power of a strong A_∞ weight. Regularity results are achieved under minimal assumptions on the coefficients.

Keywords Harnack inequality · Strong A_∞ weights · Stummel-Kato classes

1 Introduction

This paper is a contribution towards a complete regularity theory concerning solutions of degenerate elliptic equations under minimal assumptions on the coefficients. Here we consider quasilinear elliptic equations whose ellipticity degenerates as a suitable power of a strong A_∞ weight. The class of strong A_∞ weights has been introduced by David and Semmes in [2] for different purposes and it has been found useful in several problems related to geometric measure theory and quasiconformal mappings. An important case—in connection with regularity and PDEs—is indeed the case of the Jacobian of a quasiconformal mapping.

In the last decades, degenerate linear equations have been studied and some regularity results for them have been proved (see e.g. [5, 7, 9, 11, 13, 21, 22]). We remark that the above cited results are concerned with linear equations with Muckenhoupt A_2 degeneracy only.

The classes of strong A_∞ weights and A_2 are not comparable i.e. not any strong A_∞ weight is a A_2 Muckenhoupt weight and not any A_2 weight is a strong A_∞ weight.

G. Di Fazio · M. S. Fanciullo (✉) · P. Zamboni
Dipartimento di Matematica e Informatica, Università di Catania, Catania, Italy
e-mail: fanciullo@dmi.unict.it

V. Vespri et al. (eds.), *Harnack Inequalities and Nonlinear Operators*,
Springer INdAM Series 46, https://doi.org/10.1007/978-3-030-73778-8_7

The novelty in this paper is threefold. Namely, the main point is the special kind of degeneracy; then the minimal assumptions regarding the lower order terms and finally the regularity up to the boundary.

We now briefly describe the content of the paper. Let us consider quasilinear elliptic equations in divergence form

$$\operatorname{div} A(x, u, \nabla u) + B(x, u, \nabla u) = 0\,, \tag{1}$$

where A and B are measurable functions satisfying suitable structure conditions

$$\begin{cases} |A(x,u,\xi)| \le a w(x)|\xi|^{p-1} + b(x)|u|^{p-1} + e(x) \\ |B(x,u,\xi)| \le b_0 w(x)|\xi|^p + b_1(x)|\xi|^{p-1} + d(x)|u|^{p-1} + f(x) \\ \xi \cdot A(x,u,\xi) \ge w(x)|\xi|^p - d_1(x)|u|^p - g(x)\,. \end{cases} \tag{2}$$

Here $1 < p < n$, $w = v^{1-\frac{p}{n}}$, v is a strong A_∞ weight and the coefficients of the lower order terms belong to suitable Stummel–Kato or Morrey classes. We stress that the function B is required to have natural growth in the variable ξ.

Equation (1) has extensively been investigated in the case $v \equiv 1$. Here we quote some contributions—among others—by Trudinger and Lieberman. In [20] (see also [12]) Trudinger considers the same equation with no degeneracy and coefficients in suitable L^p classes. There, Harnack inequality and regularity properties of bounded weak solutions are proved. In [15] Lieberman considers equations

$$\operatorname{div} A(x, u, \nabla u) + B(x, u, \nabla u) = \mu \tag{3}$$

assuming μ to be a given signed Radon measure satisfying a Morrey type condition. There, Harnack inequality and regularity for bounded weak solutions are proved under the structure conditions (2) with lower order terms in suitable Morrey classes.

In this paper we show that Trudinger technique to prove Harnack inequality can be carried out up to the boundary for degenerate quasilinear equations with coefficients that satisfy minimal integrability assumptions (see also [6]).

More precisely, we assume the coefficients in suitable Stummel-Kato and Morrey classes and use a Fefferman type inequality proved in [4] (see also [8, 10, 17] and [18]) to control the integrals arising from the lower order terms.

2 Strong A_∞ Weights and Function Spaces

Let v be a A_∞ weight in $\mathbb{R}^n$. This means that, for any $\varepsilon > 0$ there exists $\delta > 0$ such that if Q is a cube in $\mathbb{R}^n$ and E is a measurable subset of Q for which $|E| \le \delta|Q|$, then $v(E) \le \varepsilon v(Q)$, i.e. $\int_E v(x)\,dx \le \varepsilon \int_Q v(x)\,dx$. If $v \in A_\infty$ and $B_{x,y}$ is the euclidean ball containing x and y whose diameter is $|x - y|$, we can define a quasi distance δ in $\mathbb{R}^n$ by setting

$$\delta(x, y) = \left(\int_{B_{x,y}} v(t)\, dt \right)^{1/n}.$$

We remark that $\delta(x, y) = |x - y|$ when $v(t) \equiv 1$. By using the function $\delta(x, y)$ we may define the δ-length of a curve as the limsup of the δ-lengths of the approximating polygonals.

On the other side we can actually define a distance related to the weight v. We take, as the distance between two points x and y, the infimum of the δ-length of the curves connecting x and y. Namely we set,

$$d_v(x, y) = \inf\{\delta\text{-length of the curves connecting } x \text{ and } y\}.$$

In general, the function δ is not comparable to a distance.

Definition 2.1 *If v is an A_∞ weight there exists a positive constant c such that $\delta(x, y) \leq c\, d_v(x, y)$, for any $x, y \in \mathbb{R}^n$ (see [2]). If, in addition,*

$$\delta(x, y) \sim d_v(x, y) \quad \forall x, y \in \mathbb{R}^n \tag{4}$$

we say that v is a strong A_∞ weight.

In this section we denote by $B \equiv B(x, R)$ and $B_e \equiv B(x, R)$ respectively the metric and euclidean balls centered at x with radius R. The measure $v\, dx$ is Ahlfors regular and, as a consequence, is a doubling measure (see e.g. [19]).

Theorem 2.1 *Let v be a strong A_∞ weight. Then, there exist two positive constants a and A, depending only on n and the comparability constants in (4), such that for any $x \in \mathbb{R}^n$ and any $r > 0$, we have*

$$a\, r^n \leq v(B(x, r)) \leq A\, r^n.$$

Moreover, there exists $c > 0$ such that for any $r > 0$ there exists $R = R(r)$ such that

$$B_e(x, cR) \subseteq B(x, r) \subseteq B_e(x, R) \qquad \forall x \in \mathbb{R}^n.$$

The classes of A_p Muckenhoupt and strong A_∞ weights are not comparable.

Remark 2.1 Any strong A_∞ weight is a A_∞ weight. For any $1 < p < \infty$ there exists an A_p weight which is not a strong A_∞ weight.

Now we quote Poincaré and Sobolev inequalities.

Theorem 2.2 ([14]) *Let v be a strong A_∞ weight and $1 < p < n$. Let q be such that $v \in A_q$. The following Sobolev inequality and Poincaré inequality hold true*

$$\left(\fint_B |u(x)|^{kp}\, w\, dx\right)^{\frac{1}{kp}} \le c\ \mathrm{diam}(B) \left(\fint_B |\nabla u(x)|^p\, w\, dx\right)^{\frac{1}{p}} \forall u \in C_0^\infty(B) \tag{5}$$

$$\fint_B |u - u_B|^p\, w\, dx \le c(\mathrm{diam}\, B)^p \fint_B |\nabla u|^p\, w\, dx \qquad \forall u \in C^\infty(B) \tag{6}$$

where $w(x) = v(x)^{1-\frac{p}{n}}$, the symbol $\fint$ denotes the average with respect to the measure $w(x)\, dx$, $k = \dfrac{p + q(n-p)}{q(n-p)}$ and B denotes a metric ball.

Using strong A_∞ weights we define Lebesgue and Sobolev classes.

Definition 2.2 *Let v be a strong A_∞ weight, $1 < p < n$ and $w = v^{1-p/n}$, $\Omega \subset \mathbb{R}^n$. For any $u \in C_0^\infty(\Omega)$ we set*

$$\|u\|_{p,v} = \left(\int_\Omega |u(x)|^p\, w(x)\, dx\right)^{1/p} \qquad 1 \le p < \infty. \tag{7}$$

We define $L_v^p(\Omega)$ to be the completion of $C_0^\infty(\Omega)$ with respect to the above norm. In a similar way we define Sobolev classes. For any $u \in C^\infty(\Omega)$ we set

$$\|u\|_{1,p,v} = \left(\int_\Omega |u(x)|^p\, w(x)\, dx\right)^{1/p} + \left(\int_\Omega |\nabla u(x)|^p\, w(x)\, dx\right)^{1/p}. \tag{8}$$

We define $W_{0,v}^{1,p}(\Omega)$ to be the completion of $C_0^\infty(\Omega)$ with respect to the above norm and $W_v^{1,p}(\Omega)$ to be the completion of $C^\infty(\Omega)$ with respect to the same norm.

In the above definitions we put v in the symbol of the norm and w into the integrals. This is because we want to stress the dependence on the strong A_∞ weight v.

In order to formulate the assumptions on the lower order terms we need to define some other function spaces.

Definition 2.3 *Let $1 < p < n$, f be a locally integrable function in $\Omega \subset \mathbb{R}^n$ and let v be a strong A_∞ weight. We set*

$$\phi(f; R) = \sup_{x \in \Omega} \left(\int_{B(x,R)} k(x,y) \left(\int_{B(x,R)} |f(z)| k(z,y)\, v(z)^{1-\frac{p}{n}}\, dz\right)^{\frac{1}{p-1}} v(y)\, dy\right)^{p-1} \tag{9}$$

where

$$k(x, y) = \frac{1}{v(B(x, d_v(x, y)))^{1-\frac{1}{n}}}.$$

We shall say that f belongs to the class $\tilde{S}_v(\Omega)$ if $\phi(f; R)$ is bounded in a neighborhood of the origin. Moreover, if $\lim_{R\to 0} \phi(f; R) = 0$ then we say that f belongs to the Stummel-Kato class $S_v(\Omega)$. If there exists $\rho > 0$ such that

$$\int_0^\rho \frac{\phi(f; t)^{1/p}}{t} dt < +\infty, \tag{10}$$

then we say that the function f belongs to the class $S_v^{'}(\Omega)$.

Definition 2.4 (Morrey Spaces) *Let $1 \le q < n$, $1 < p < n$ and v be a strong A_∞ weight. We say that f belongs to $L_v^{q,\lambda}(\Omega)$, for some $\lambda > 0$, if*

$$\|f\|_{L_v^{q,\lambda}(\Omega)} =$$

$$= \sup_{x\in\Omega, 0<r<d_0} \left(\frac{r^\lambda}{|B(x,r)\cap\Omega|} \int_{B(x,r)\cap\Omega} |f(y)|^q v(y)^{1-\frac{p}{n}} dy \right)^{\frac{1}{q}} < +\infty,$$

where $d_0 = \operatorname{diam}(\Omega)$.

Remark 2.2 It is an easy task to check that the above definitions give back their classical counterparts when $v \equiv 1$.

It is easy to compare the function classes previously defined.

Proposition 2.1 *Let p and ε be numbers such that $1 < p < n$ and $0 < \varepsilon < p$. The class $L_v^{1,p-\varepsilon}(\Omega)$ is embedded in $S_v^{'}(\Omega)$. Namely, if $V \in L_v^{1,p-\varepsilon}(\Omega)$ then*

$$\phi(V; r) \le C\|V\|_{L_v^{1,p-\varepsilon}} r^{\frac{\varepsilon}{p-1}}$$

for any $0 < r < \operatorname{diam}\Omega$.

Proof The proof is standard and can be easily adapted from the case $w = 1$ (see [3]). □

The following lemma will be useful in the iteration process.

Lemma 2.1 ([16]) *Let $p > 1$, $\mu(r)$ a continuous positive increasing function defined in $]0, +\infty[$ such that $\lim_{r\to 0} \mu(r) = 0$, $0 < \theta < 1$. The series*

$$\sum_{j=0}^{+\infty} \theta^j \log \mu^{-1}\left(\theta^{pj}\right)$$

is convergent if and only if there exists $\rho > 0$ such that

$$\int_0^\rho \frac{\mu^{\frac{1}{p}}(t)}{t} dt < +\infty.$$

The following result will be useful in the proof of the weak Harnack inequality.

Theorem 2.3 ([4]) *Let Ω be a bounded domain in $\mathbb{R}^n$ and let V belong to the class $\tilde{S}_v(\Omega)$. If v is a strong A_∞ weight and $1 < p < n$, then there exists a constant c such that for any $u \in C_0^\infty(\Omega)$ we have*

$$\left(\int_B |V(x)||u(x)|^p\, w\, dx\right)^{1/p} \leq c\,\phi^{1/p}(V; 2R) \left(\int_B |\nabla u(x)|^p w\, dx\right)^{1/p} \tag{11}$$

where $w(x) \equiv v^{1-\frac{p}{n}}(x)$ and R is the radius of a metric ball $B \equiv B_R$, containing the support of u.

As a direct consequence we have

Corollary 2.1 *Let $1 < p < n$ and v be a strong A_∞ weight. Let V belongs to the class $S_v(\Omega)$. Then, for any $\varepsilon > 0$, there exists $K(\varepsilon)$ such that*

$$\int_\Omega |V(x)||u(x)|^p w(x)\, dx \leq \varepsilon \int_\Omega |\nabla u(x)|^p w(x)\, dx +$$

$$+ K(\varepsilon) \int_\Omega |u(x)|^p w(x)\, dx \;\; \forall u \in C_0^\infty(\Omega) \tag{12}$$

where $w(x) = v(x)^{1-\frac{p}{n}}$, $K(\varepsilon) \sim \dfrac{\sigma}{\left[\phi^{-1}(V; \varepsilon)\right]^{n+p}}$ and ϕ^{-1} denotes the inverse function of ϕ.

3 Harnack Inequality

In this section we prove a weak Harnack inequality for non negative weak solutions of the equation

$$\mathrm{div} A(x, u, \nabla u) + B(x, u, \nabla u) = 0 . \tag{13}$$

We recall what we mean by weak solution of (13).

Definition 3.1 *A function $u \in W_v^{1,p}(\Omega)$ is a local weak subsolution (supersolution) of Eq. (13) in Ω if*

$$\int_\Omega A(x, u(x), \nabla u(x)) \cdot \nabla\varphi \, dx - \int_\Omega B(x, u(x), \nabla u(x))\varphi \, dx \leq 0 \ (\geq 0) \tag{14}$$

for every non negative $\varphi \in W_{0,v}^{1,p}(\Omega)$. A function u is a weak solution if it is both super and sub solution.

We require the functions $A(x, u, p)$ and $B(x, u, p)$ to be measurable functions satisfying the following structure conditions

$$\begin{cases} |A(x, u, \xi)| \leq a w(x)|\xi|^{p-1} + b(x)|u|^{p-1} \\ |B(x, u, \xi)| \leq b_0 w(x)|\xi|^p + b_1(x)|\xi|^{p-1} + d(x)|u|^{p-1} \\ \xi \cdot A(x, u, \xi) \geq w(x)|\xi|^p - d_1(x)|u|^p \end{cases} \tag{15}$$

where $1 < p < n$, $w = v^{1-\frac{p}{n}}$ and v is a strong A_∞ weight.

We show that locally bounded weak solutions verify a Harnack inequality and, as a consequence, some regularity properties. We shall make the following assumptions on the lower order terms to ensure the continuity of local weak solutions

$$a, b_0 \in \mathbb{R}, \left(\frac{b}{w}\right)^{\frac{p}{p-1}}, \left(\frac{b_1}{w}\right)^p, \frac{d}{w}, \frac{d_1}{w} \in S_v'(\Omega) . \tag{16}$$

From now on we denote by $B_r = B_r(x)$ the euclidean ball centered at x with radius r.

Theorem 3.1 *Let u be a non negative weak supersolution of Eq. (13) in Ω satisfying (15) and (16). Let B_r be a ball such that $B_{3r} \Subset \Omega$ and let M be a constant such that $u \leq M$ in B_{3r}. Then there exists c depending on n, M, a, b_0, p and the weight v such that*

$$w^{-1}(B_{2r}) \int_{B_{2r}} u \, w \, dx \leq c \inf_{B_r} u .$$

Proof We can suppose $u > 0$ in Ω.[1] We take $\varphi = \eta^p u^\beta e^{-b_0 u}$, $\beta < 0$ as test function in (14) so we obtain

$$\int_{B_{3r}} \eta^p e^{-b_0 u}(b_0 u^\beta + |\beta| u^{\beta-1}) \nabla u \cdot A dx - \\ p \int_{B_{3r}} u^\beta \eta^{p-1} e^{-b_0 u} \nabla \eta \cdot A dx + \int_{B_{3r}} \eta^p u^\beta e^{-b_0 u} B dx \leq 0 .$$

The previous inequality and the structure assumptions (15) yield

$$\int_{B_{3r}} e^{-b_0 u} \eta^p (b_0 u^\beta + |\beta| u^{\beta-1}) |\nabla u|^p w dx \leq \\ \int_{B_{3r}} e^{-b_0 u} \eta^p (b_0 u^\beta + |\beta| u^{\beta-1})(\nabla u \cdot A + d_1 |u|^p) dx \leq \\ p \int_{B_{3r}} u^\beta \eta^{p-1} e^{-b_0 u} \nabla \eta \cdot A\, dx - \int_{B_{3r}} \eta^p u^\beta e^{-b_0 u} B\, dx + \\ \int_{B_{3r}} e^{-b_0 u} \eta^p (b_0 u^\beta + |\beta| u^{\beta-1}) d_1 |u|^p\, dx \leq \\ p \int_{B_{3r}} u^\beta \eta^{p-1} e^{-b_0 u} \nabla \eta \cdot A\, dx + \\ \int_{B_{3r}} \eta^p u^\beta e^{-b_0 u} (b_0 |\nabla u|^p w + b_1 |\nabla u|^{p-1} + d |u|^{p-1}) dx + \\ \int_{B_{3r}} e^{-\beta b_0 u} \eta^p (b_0 u^\beta + |\beta| u^{\beta-1}) d_1 |u|^p\, dx .$$

By boundedness of u in B_{3r} we obtain

$$|\beta| \int_{B_{3r}} \eta^p u^{\beta-1} |\nabla u|^p w dx \leq cp \int_{B_{3r}} u^\beta \eta^{p-1} \nabla \eta \cdot A\, dx + \\ c \int_{B_{3r}} \eta^p u^\beta (b_1 |\nabla u|^{p-1} + d |u|^{p-1}) dx + \\ c \int_{B_{3r}} \eta^p (b_0 u^\beta + |\beta| u^{\beta-1}) d_1 |u|^p\, dx \leq \\ c \int_{B_{3r}} \Big\{ u^\beta \eta^{p-1} |\nabla \eta| (a w |\nabla u|^{p-1} + b |u|^{p-1}) +$$

[1] We should take $u + \epsilon$ ($\epsilon > 0$) which is positive in Ω and, after obtaining estimates independing of ϵ, go to the limit for $\epsilon \to 0$.

$$\eta^p u^\beta b_1|\nabla u|^{p-1} + \eta^p u^{\beta+p-1} d +$$

$$+ \eta^p b_0 u^{\beta+p} d_1 + |\beta|\eta^p u^{\beta+p-1} d_1 \Big\} dx \leq$$

$$c \int_{B_{3r}} \Big\{ a u^\beta \eta^{p-1} |\nabla \eta||\nabla u|^{p-1} w + u^{\beta+p-1}\eta^{p-1}|\nabla\eta| b +$$

$$\eta^p u^\beta b_1 |\nabla u|^{p-1} + (1+|\beta|)\eta^p u^{\beta+p-1}(d+d_1) + \eta^p b_0 u^{\beta+p} d_1 \Big\} dx \,.$$

Then, from Young inequality

$$|\beta| \int_{B_{3r}} \eta^p u^{\beta-1} |\nabla u|^p w dx \leq$$

$$\leq c(a, b_0, M, p) \int_{B_{3r}} \Big\{ u^\beta \eta^{p-1} |\nabla\eta||\nabla u|^{p-1} w \, dx +$$

$$+ u^{\beta+p-1}|\nabla \eta|^p w + \eta^p u^{\beta+p-1} \frac{b^{\frac{p}{p-1}}}{w^{\frac{1}{p-1}}} +$$

$$+ \epsilon \eta^p u^{\beta-1} |\nabla u|^p w + c(\epsilon)\eta^p \frac{b_1^p}{w^{p-1}} u^{\beta+p-1} +$$

$$+ (1+|\beta|)\eta^p u^{\beta+p-1}(d+d_1) \Big\} dx \,.$$

We set $V = \frac{b^{\frac{p}{p-1}}}{w^{\frac{1}{p-1}}} + \frac{b_1^p}{w^{p-1}} + d + d_1$ in order to get short the previous inequality. We obtain

$$\int_{B_{3r}} \eta^p u^{\beta-1} |\nabla u|^p w dx \leq$$

$$\leq c(1+|\beta|^{-1})^p \int_{B_{3r}} \Big\{ |\nabla\eta|^p u^{\beta+p-1} w + V\eta^p u^{\beta+p-1} \Big\} dx \,. \tag{17}$$

Now the proof follows the lines of Theorem 4.3 in [4]. We set

$$\mathcal{U}(x) = \begin{cases} u^q(x) & \text{where} \quad pq = p+\beta-1 \quad \text{if} \quad \beta \neq 1-p \\ \log u(x) & \text{if} \quad \beta = 1-p \end{cases}$$

by (17) we have

$$\int_{B_{3r}} \eta^p |\nabla \mathcal{U}|^p w\, dx \le c|q|^p(1+|\beta|^{-1})^p \left\{ \int_{B_{3r}} |\nabla \eta|^p \mathcal{U}^p w\, dx + \right.$$

$$\left. + \int_{B_{3r}} V \eta^p \mathcal{U}^p\, dx \right\}, \beta \neq 1-p \tag{18}$$

while

$$\int_{B_{3r}} \eta^p |\nabla \mathcal{U}|^p w\, dx \le c \left\{ \int_{B_{3r}} |\nabla \eta|^p w\, dx + \int_{B_{3r}} V \eta^p\, dx \right\} \tag{19}$$

if $\beta = 1-p$. Let us start with the case $\beta = 1-p$. By Theorem 2.3 we have

$$\int_{B_{3r}} V \eta^p\, dx \le c\phi\left(\frac{V}{w}; \operatorname{diam} \Omega\right) \int_{B_{3r}} |\nabla \eta|^p w\, dx\,,$$

and from (19)

$$\int_{B_{3r}} \eta^p |\nabla \mathcal{U}|^p w\, dx \le c \int_{B_{3r}} |\nabla \eta|^p w\, dx\,.$$

Let B_h be a ball contained in B_{2r}. Choosing a smooth function η so that $\eta = 1$ in B_h, $0 \le \eta \le 1$ in $B_{2r} \setminus B_h$ and $|\nabla \eta| \le \frac{c}{h}$, we get

$$\|\nabla \mathcal{U}\|_{L^p_v(B_h)} \le c \frac{w(B_h)^{\frac{1}{p}}}{h}\,.$$

By Poincaré inequality (6) and John–Nirenberg lemma (see [1]) we have $\mathcal{U}(x) = \log u(x) \in BMO_v$. Then there exist two positive constants p_0 and c, such that

$$\left(\fint_{B_{2r}} e^{p_0 \mathcal{U}} w\, dx \right)^{\frac{1}{p_0}} \left(\fint_{B_{2r}} e^{-p_0 \mathcal{U}} w\, dx \right)^{\frac{1}{p_0}} \le c\,. \tag{20}$$

Let us consider the following family of seminorms

$$\Phi(p,h) = \left(\int_{B_h} |u|^p w\, dx \right)^{1/p}, \quad p \neq 0\,.$$

By (20) we have

$$\frac{1}{w(B_{2r})^{1/p_0}} \Phi(p_0, 2r) \le c w(B_{2r})^{1/p_0} \Phi(-p_0, 2r)\,. \tag{21}$$

In the case (18) by Corollary 2.1 we obtain

$$\int_{B_{3r}} |\nabla \mathcal{U}|^p \eta^p w\,dx \le c\left\{(|q|^p+1)\left(1+\frac{1}{|\beta|}\right)^p \int_{B_{3r}} |\nabla \eta|^p \mathcal{U}^p w\,dx + \right.$$
$$\left. + \left[\frac{1}{\phi^{-1}\left(\frac{V}{w}; |q|^{-p}\left(1+\frac{1}{|\beta|}\right)^{-p}\right)}\right]^{n+p} \int_{B_{3r}} \eta^p \mathcal{U}^p w\,dx \right\}. \tag{22}$$

By Sobolev inequality we have

$$\left(\int_{B_{3r}} |\eta \mathcal{U}|^{kp} w\,dx\right)^{\frac{1}{k}} \le c w(B)^{\frac{1}{k}-1}\left\{(|q|^p+2)\left(1+\frac{1}{|\beta|}\right)^p \cdot \right.$$
$$\int_{B_{3r}} |\nabla \eta|^p \mathcal{U}^p w\,dx +$$
$$\left. + \left[\frac{1}{\phi^{-1}\left(\frac{V}{w}; |q|^{-p}\left(1+\frac{1}{|\beta|}\right)^{-p}\right)}\right]^{n+p} \int_{B_{3r}} \eta^p \mathcal{U}^p w\,dx \right\} \tag{23}$$

where c is a positive constant independent of w.

Now we choose the function η. Let r_1 and r_2 be real numbers such that $r \le r_1 < r_2 \le 2r$ and let the function η be chosen so that $\eta(x) = 1$ in B_{r_1}, $0 \le \eta(x) \le 1$ in B_{r_2}, $\eta(x) = 0$ outside B_{r_2}, $|\nabla \eta| \le \frac{c}{r_2 - r_1}$ for some fixed constant c. We have

$$\left(\int_{B_{r_1}} \mathcal{U}^{kp} w\,dx\right)^{\frac{1}{k}} \le c w(B)^{\frac{1}{k}-1} \frac{1}{(r_2-r_1)^p}(|q|^p+2) \cdot$$
$$\cdot \left(1+\frac{1}{|\beta|}\right)^p \left[\frac{1}{\phi^{-1}\left(\frac{V}{w}; |q|^{-p}\left(1+\frac{1}{|\beta|}\right)^{-p}\right)}\right]^{n+p} \int_{B_{r_2}} \mathcal{U}^p w\,dx\,.$$

Setting $\gamma = pq = p + \beta - 1$ and recalling that $\mathcal{U}(x) = u^q(x)$, we get

$$\Phi(k\gamma, r_1) \ge c^{\frac{1}{\gamma}} w(B)^{\frac{1}{\gamma}(\frac{1}{k}-1)} (|q|^p+2)^{\frac{1}{\gamma}} \cdot$$
$$\cdot \left[\frac{1}{\phi^{-1}\left(\frac{V}{w}; |q|^{-p}\right)}\right]^{\frac{n+p}{\gamma}} \frac{1}{(r_2-r_1)^{\frac{1}{p}}} \Phi(\gamma, r_2)\,, \tag{24}$$

for negative γ. This is the inequality we are going to iterate. If $\gamma_i = k^i p_0$ and $r_i = r + \frac{r}{2^i}, i = 1, 2, \ldots$ iteration of (24) and use of Lemma 2.1 yield

$$\Phi(-\infty, r) \geq c(p, a, \phi_{\frac{V}{w}}, \operatorname{diam}\Omega) w(B_r)^{\frac{1}{p_0}} \Phi(-p_0, 2r) .$$

Therefore from (21) and Hölder inequality we obtain

$$w^{-1}(B_{2r})\Phi(1, 2r) \leq c\Phi(-\infty, r)$$

where $c \equiv c(p, a, \phi_{\frac{V}{w}}, \operatorname{diam}\Omega)$ and the result follows. □

We obtain a weak Harnack inequality for weak subsolutions in a similar way of Theorem 3.1.

Theorem 3.2 *Let u be a non negative weak subsolution of Eq.* (13) *in Ω satisfying* (15) *and* (16)*. Let B_r be a ball such that $B_{3r} \Subset \Omega$ and let M be a constant such that $u \leq M$ in B_{3r}. Then there exists c depending on n, M, a, b_0, p and the weight v such that*

$$\sup_{B_r} u \leq c w^{-1}(B_{2r}) \int_{B_{2r}} u\, w dx$$

If we take a non negative weak solution, we can put together the two previous results.

Theorem 3.3 *Let u be a non negative weak solution of Eq.* (13) *in Ω satisfying* (15) *and* (16)*. Let B_r be a ball such that $B_{3r} \Subset \Omega$ and let M be a constant such that $u \leq M$ in B_{3r}. Then there exists c depending on n, M, a, b_0, p and the weight v such that*

$$\sup_{B_r} u \leq c \inf_{B_r} u .$$

Now, as a simple consequence of Harnack inequality, we get some regularity results for weak solutions of (13). The proof is an immediate consequence of Harnack inequality so we omit it.

Theorem 3.4 *Let u be a locally bounded weak solution of Eq.* (13) *in Ω satisfying* (15) *and* (16)*. Then u is continuous in Ω.*

If we assume more restrictive assumptions on the lower order terms we obtain the following refinement of the previous one.

Theorem 3.5 *Let u be a locally bounded weak solution of Eq. (13) in Ω satisfying (15) and*

$$a, b_0 \in \mathbb{R}, \left(\frac{b}{w}\right)^{\frac{p}{p-1}}, \left(\frac{b_1}{w}\right)^p, \frac{d}{w}, \frac{d_1}{w} \in L_v^{1,p-\varepsilon}(\Omega), \quad \varepsilon > 0.$$

Then u is locally Hölder continuous in Ω.

4 Boundary Harnack Inequality

Our next step is to show a Harnack inequality near the boundary of Ω for weak supersolutions and subsolutions to the equation

$$\mathrm{div} A(x, u, \nabla u) + B(x, u, \nabla u) = 0, \tag{25}$$

with the following structure conditions

$$\begin{cases} |A(x, u, \xi)| \leq a w(x)|\xi|^{p-1} + e(x) \\ |B(x, u, \xi)| \leq b_0 w(x)|\xi|^p + b_1(x)|\xi|^{p-1} + f(x) \\ \xi \cdot A(x, u, \xi) \geq w(x)|\xi|^p - g(x) \end{cases} \tag{26}$$

where $1 < p < n$, $w = v^{1-\frac{p}{n}}$ and v is a strong A_∞ weight and

$$a, b_0 \in \mathbb{R}, \left(\frac{b_1}{w}\right)^p, \left(\frac{e}{w}\right)^{\frac{p}{p-1}}, \frac{f}{w}, \frac{g}{w} \in S_v'(\Omega). \tag{27}$$

Let B_r be a ball centered in $x_0 \in \partial\Omega$. If u is a weak supersolution of (25) we set

$$\tilde{u}(x) = \begin{cases} \min\{u, m\} & \text{if } x \in \Omega \cap B_{4r} \\ m & \text{if } x \in \mathbb{R}^n \setminus (\Omega \cap B_{4r}) \end{cases}$$

where $m = \inf_{\partial\Omega \cap B_{4r}} u$.

Theorem 4.1 *Let $u \in W_v^{1,p}(\Omega \cap B_{4r})$ be a weak non negative supersolution of (25) in $\Omega \cap B_{4r}$. Assume (26) and (27). Let M be a constant such that $u \leq M$ on $\Omega \cap B_{4r}$. Then there exists c depending on n, M, a, b_0, p and the weight v such that*

$$w^{-1}(B_{2r}) \int_{B_{2r}} \tilde{u}\, w dx \leq$$
$$\leq c \left\{ \inf_{B_r} \tilde{u} + \phi^{1/p}\left[\left(\frac{e}{w}\right)^{\frac{p}{p-1}}; r\right] + \phi^{\frac{1}{p-1}}\left(\frac{f}{w}; r\right) + \phi^{\frac{1}{p}}\left(\frac{g}{w}; r\right) \right\}. \tag{28}$$

Proof We define $b_1 = 0$, $e = 0$, $f = 0$ and $g = 0$ outside Ω. Set $k = \phi^{1/p}\left[\left(\frac{e}{w}\right)^{\frac{p}{p-1}}; r\right] + \phi^{\frac{1}{p-1}}\left(\frac{f}{w}; r\right) + \phi^{\frac{1}{p}}\left(\frac{g}{w}; r\right)$ and $v = \tilde{u} + k$. From (26) we get

$$\begin{cases} |A(x, u, \nabla u)| \leq a w(x)|\nabla v|^{p-1} + b(x)|v|^{p-1} \\ |B(x, u, \nabla u)| \leq b_0 w(x)|\nabla v|^p + b_1(x)|\nabla v|^{p-1} + d(x)|v|^{p-1} \\ \xi \cdot A(x, u, \nabla u) \geq w(x)|\nabla v|^p - d_1(x)|v|^p \end{cases} \tag{29}$$

where $b(x) = \frac{e(x)}{k^{p-1}}$, $d(x) = \frac{f(x)}{k^{p-1}}$ and $d_1(x) = \frac{g(x)}{k^p}$.

Since for any $0 < \rho < 4r$

$$\phi\left(\left(\frac{b}{w}\right)^{\frac{p}{p-1}}; \rho\right) = \frac{1}{k^p}\phi\left(\left(\frac{e}{w}\right)^{\frac{p}{p-1}}; \rho\right)$$

$$\phi\left(\frac{d}{w}; \rho\right) = \frac{1}{k^{p-1}}\phi\left(\frac{f}{w}; \rho\right)$$

$$\phi\left(\frac{d_1}{w}; \rho\right) = \frac{1}{k^p}\phi\left(\frac{g}{w}; \rho\right)$$

we get

$$\left(\frac{b}{w}\right)^{\frac{p}{p-1}}, \frac{d}{w} \text{ and } \frac{d_1}{w} \in S_v'(B_{3r}).$$

Let $\eta \in C_0^1(B_{3r})$ and $\eta \geq 0$. For $\beta < 0$ we take $\varphi(x) = \eta^p[v^\beta - (m + k)^\beta]e^{-|b_0|v(x)} \in W_{0,v}^{1,p}(B_{4r})$ as test function. Since u is a supersolution of (25) we have

$$\int_{B_{4r}} \eta^p e^{-|b_0|v}\{|\beta|v^{\beta-1} + b_0[v^\beta - (m+k)^\beta]\}A(x, u, \nabla u) \cdot \nabla v dx +$$

$$+ \int_{B_{4r}} B(x, u, \nabla u)\varphi dx \leq$$

$$\leq p \int_{B_{4r}} \eta^{p-1} A(x, u, \nabla u) \cdot \nabla\eta[v^\beta - (m+k)^\beta]e^{-|b_0|v} dx \tag{30}$$

Using (29) we get

$$\int_{B_{4r}} (w|\nabla v|^p d_1 v^p)\eta^p e^{-|b_0|v}\{|\beta|v^{\beta-1} + b_0[v^\beta - (m+k)^\beta]\}dx \leq$$

$$\leq p\int_{B_{4r}} (aw|\nabla v|^{p-1} + b_0 v^{p-1})\eta^{p-1}|\nabla\eta|[v^\beta - (m+k)^\beta]e^{-|b_0|v}+$$

$$+\int_{B_{4r}} (b_0 w|\nabla v|^p + b_1|\nabla v|^{p-1} + dv^{p-1})\eta^p[v^\beta - (m+k)^\beta]e^{-|b_0|v(x)} \tag{31}$$

from which

$$\int_{B_{4r}} |\beta||\nabla v|^p\eta^p e^{-|b_0|v}v^{\beta-1}wdx \leq |\beta|\int_{B_3r} d_1\eta^p v^{p+\beta-1}e^{-|b_0|v}dx+$$

$$+ b_0\int_{B_{4r}} d_1\eta^p v^p[v^\beta - (m+k)^\beta]e^{-|b_0|v(x)}+$$

$$+ p\int_{B_{4r}} a\eta^{p-1}|\nabla\eta||\nabla v|^{p-1}[v^\beta - (m+k)^\beta]e^{-|b_0|v}wdx+$$

$$+ p\int_{B_{4r}} bv^{p-1}\eta^{p-1}|\nabla\eta|[v^\beta - (m+k)^\beta]e^{-|b_0|v}dx+$$

$$+\int_{B_{4r}} b_1\eta^p|\nabla v|^{p-1}[v^\beta - (m+k)^\beta]e^{-|b_0|v}dx+$$

$$+\int_{B_{4r}} dv^{p-1}\eta^p[v^\beta - (m+k)^\beta]e^{-|b_0|v}dx\,. \tag{32}$$

Then

$$\int_{B_{4r}} |\beta||\nabla v|^p\eta^p v^{\beta-1}wdx \leq c|\beta|\int_{B_3r} d_1\eta^p v^{p+\beta-1}dx+$$

$$+ c\int_{B_{4r}} d_1\eta^p v^p[v^\beta - (m+k)^\beta]+$$

$$+ c\int_{B_{4r}} a\eta^{p-1}|\nabla\eta||\nabla v|^{p-1}[v^\beta - (m+k)^\beta]wdx+$$

$$+ c\int_{B_{4r}} bv^{p-1}\eta^{p-1}|\nabla\eta|[v^\beta - (m+k)^\beta]dx+$$

$$+ c\int_{B_{4r}} b_1\eta^p|\nabla v|^{p-1}[v^\beta - (m+k)^\beta]dx+$$

$$+ c\int_{B_{4r}} dv^{p-1}\eta^p[v^\beta - (m+k)^\beta]dx\,. \tag{33}$$

Since $v^\beta - (m+k)^\beta \leq v^\beta$ the proof follows as the proof of Theorem 3.1. □

Now, let u be a subsolution of (25). We define the function

$$\overline{u}(x) = \begin{cases} \max\{u, M\} & \text{if } x \in \Omega \cap B_{4r} \\ M & \text{if } x \in \mathbb{R}^n \setminus (\Omega \cap B_{4r}) \end{cases}$$

where $M = \sup_{\partial\Omega \cap B_{4r}} u$.

Theorem 4.2 *Let $u \in W_v^{1,p}(\Omega \cap B_{4r})$ be a weak non negative subsolution of (25) in $\Omega \cap B_{4r}$. Assume (26) and (27). Let M be a constant such that $u \leq M$ on $\Omega \cap B_{4r}$. Then there exists c depending on n, M, a, b_0, p and the weight v such that*

$$\sup_{B_r} \overline{u} \leq c \Bigg\{ w^{-1}(B_{2r}) \int_{B_{2r}} \overline{u}\, w\, dx + \\ + \phi^{1/p}\left[\left(\frac{e}{w}\right)^{\frac{p}{p-1}}; r\right] + \phi^{\frac{1}{p-1}}\left(\frac{f}{w}; r\right) + \phi^{\frac{1}{p}}\left(\frac{g}{w}; r\right)\Bigg\}.$$

In order to get regularity up to the boundary of the domain we need some geometric assumptions.

Definition 4.1 *Let Ω be a domain in $\mathbb{R}^n$ and $x_0 \in \partial\Omega$. Let v be a strong A_∞ weight and $w = v^{1-\frac{p}{n}}$, $1 < p < n$. We say that Ω satisfies the condition A_v at x_0 if there exist positive constants R_0 and A such that*

$$\frac{w(B_r(x_0) \setminus \Omega)}{w(B_r(x_0))} \geq A \quad 0 < r < R_0 .$$

We say that Ω satisfies the condition A_v if it satisfies the condition at any point.

In the case $v = 1$ the A_v condition gives back the outer sphere condition.

Using the geometric assumption A_v we give an estimate for the oscillation of solutions near the boundary.

Theorem 4.3 *Let Ω be a bounded open set satisfying the A_v condition at $x_0 \in \partial\Omega$. Let u be a locally bounded weak solution of Eq. (25) in Ω satisfying (26) and (27). Then there exists $R_0 > 0$ such that for any ball $B_r(x_0)$, with $0 < r < R_0$ and $\mu \in (0, 1)$ we have*

$$\operatorname*{osc}_{B_r \cap \Omega} u \leq c \left[\left(\frac{r}{R_0}\right)^{\alpha} \operatorname*{osc}_{B_{R_0} \cap \Omega} u + \operatorname*{osc}_{B_{r^\mu R_0^{1-\mu}} \cap \partial\Omega} u + k(r^\mu R_0^{1-\mu}) \right],$$

with c and α positive constant, and $k(r) = \phi^{1/p}\left[\left(\frac{e}{w}\right)^{\frac{p}{p-1}}; r\right] + \phi^{\frac{1}{p-1}}\left(\frac{f}{w}; r\right) + \phi^{\frac{1}{p}}\left(\frac{g}{w}; r\right)$.

Proof For $\rho > 0$ set $M(\rho) = \sup_{B_\rho \cap \Omega} u$ and $m(\rho) = \inf_{B_\rho \cap \Omega} u$, with $B_\rho = B_\rho(x_0)$. Let $0 < r \le R_0/4$ the function $M(4r) - u$ is solution of

$$\mathrm{div}\tilde{A}(x, \overline{u}, \nabla\overline{u}) = \tilde{B}(x, \overline{u}, \nabla\overline{u}),$$

where $\tilde{A}$ and $\tilde{B}$ are defined by

$$\tilde{A}(x, \overline{u}, \overline{\xi}) = A(x, M(4r) - \overline{u}, -\overline{\xi}),$$

$$\tilde{B}(x, \overline{u}, \overline{\xi}) = B(x, M(4r) - \overline{u}, -\overline{\xi})$$

and they satisfy

$$\begin{cases} |\tilde{A}(x, \overline{u}, \overline{\xi})| \le a w(x) |\overline{\xi}|^{p-1} + e(x) \\ |\tilde{B}(x, \overline{u}, \overline{\xi})| \le b_0 w(x) |\overline{\xi}|^p + b_1(x) |\overline{\xi}|^{p-1} + f(x) \\ \overline{\xi} \cdot \tilde{A}(x, \overline{u}, \overline{\xi}) \ge w(x) |\overline{\xi}|^p - g(x). \end{cases}$$

Then by (28) and A_v condition, we have

$$\begin{aligned} M(4r) - M &\le \frac{w(B_{2r} \setminus \Omega)}{A w(B_{2r})} [M(4r) - M] = \\ &= \frac{1}{A w(B_{2r})} \int_{B_{2r} \setminus \Omega} [M(4r) - M] w dx \le \\ &\le \frac{1}{A w(B_{2r})} \int_{B_{2r} \setminus \Omega} [\widetilde{M(4r) - u}] w dx \le \\ &\le c \inf_{B_r \cap \Omega} (M(4r) - u) + k(r) \le c[M(4r) - M(r)] + k(r), \end{aligned} \tag{34}$$

where $M = \sup_{B_{4r} \cap \partial\Omega} u$ and $m = \inf_{B_{4r} \cap \partial\Omega} u$. In a similar way, taking into account the solution $u - m(4r)$, we obtain

$$m - m(4r) \le c[m(r) - m(4r)] + k(r). \tag{35}$$

By addition of (34) and (35) we obtain, for $\theta < 1$

$$M(r) - m(r) \le \theta[M(4r) - m(4r)] + M - m + ck(r),$$

from which, applying Lemma 8.23 in [20] we obtain the thesis. □

As consequences of the previous Theorem we obtain the following corollary.

Corollary 4.1 *Let Ω be a bounded open set satisfying the A_v condition in every $x_0 \in \partial\Omega$. Let u be a locally bounded weak solution of Eq. (25) in Ω satisfying (26) and (27). Let $u = \varphi$ on $\partial\Omega$. If φ is continuous in $\partial\Omega$ then u is continuous in $\overline{\Omega}$.*

Now we refine our assumptions on lower order terms. If we assume the coefficients in a suitable Morrey space we obtain Hölder continuity of the solution thanks to Proposition 2.1.

Corollary 4.2 *Let Ω be a bounded open set satisfying the A_υ condition in every $x_0 \in \partial\Omega$. Let u be a locally bounded weak solution of Eq.* (25) *in Ω satisfying* (26) *and*

$$a, b_0 \in \mathbb{R}, \left(\frac{b_1}{w}\right)^p, \left(\frac{e}{w}\right)^{\frac{p}{p-1}}, \frac{f}{w}, \frac{g}{w} \in L_\upsilon^{1,p-\epsilon}(\Omega), \tag{36}$$

with $0 < \epsilon < p$. Let $u = \varphi$ on $\partial\Omega$. If φ is Hölder continuous in $\partial\Omega$ then u is Hölder continuous in $\overline{\Omega}$.

References

1. S.M. Buckley, Inequalities of John–Nirenberg type in doubling spaces. J. Anal. Math. **79**, 215–240 (1999)
2. G. David, S. Semmes, *Strong A_∞ Weights, Sobolev Inequalities and Quasiconformal Mappings, Analysis and Partial Differential Equations*. Lecture Notes in Pure and Applied Mathematics, vol. 122 (Marcel Dekker, New York, 1990)
3. G. Di Fazio, Hölder continuity of solutions for some Schrödinger equations. Rend. Sem. Mat. Univ. Padova **79**, 173–183 (1988)
4. G. Di Fazio, P. Zamboni, Regularity for quasilinear degenerate elliptic equations. Math. Z. **253**, 787–803 (2006)
5. G. Di Fazio, M.S. Fanciullo, P. Zamboni, Harnack inequality and smoothness for quasilinear degenerate elliptic equations. J. Differential Equations **245**(10), 2939–2957 (2008)
6. G. Di Fazio, M.S. Fanciullo, P. Zamboni, Harnack inequality and regularity for degenerate quasilinear elliptic equations. Math. Z. **264**(3), 679–695 (2010)
7. G. Di Fazio, M.S. Fanciullo, P. Zamboni, Harnack inequality for a class of strongly degenerate elliptic operators formed by Hörmander vector fields. Manuscripta Math. **135**(3–4), 361–380 (2011)
8. G. Di Fazio, M.S. Fanciullo, P. Zamboni, Regularity for a class of strongly degenerate quasilinear operators. J. Differential Equations **255**(11), 3920–3939 (2013)
9. G. Di Fazio, M.S. Fanciullo, P. Zamboni, *Harnack inequality and smoothness for some non linear degenerate elliptic equations*, in *Minimax Theory and its Applications, Conference "Nonlinear Phenomena: Theory and Applications"*, vol. 4, No. 1 (2019), pp. 87–99
10. G. Di Fazio, M.S. Fanciullo, P. Zamboni, Harnack inequality and continuity of weak solutions for doubly degenerate elliptic equations. Mathematische Z. **292**, 1325–1336 (2019)
11. E. Fabes, C. Kenig, R. Serapioni, The local regularity of solutions of degenerate elliptic equations. Commun. PDE **7**(1), 77–116 (1982)
12. D. Gilbarg, N.S. Trudinger, *Elliptic Partial Differential Equations of Second Order* (Springer, Berlin, 1983)
13. C.E. Gutierrez, Harnack's inequality for degenerate Schrödinger operators. TAMS **312**, 403–419 (1989)
14. J. Heinonen, P. Koskela, Weighted Sobolev and Poincaré inequalities and quasiregular mappings of polynomial type. Math. Scand. **77**, 251–271 (1995)
15. G.M. Lieberman, Sharp form of estimates for subsolutions and supersolutions of quasilinear elliptic equations involving measures. Commun. PDE **18**, 1191–1212 (1993)

16. M.A. Ragusa, P. Zamboni, Local regularity of solutions to quasilinear elliptic equations with general structure. Commun. Appl. Anal. **3**(1), 131–147 (1999)
17. Y. Sawano, G. Di Fazio, D. Hakim, *Morrey Spaces: Introduction and Applications to Integral Operators and PDE's*, vol. I (Taylor and Francis, New York, 2020)
18. Y. Sawano, G. Di Fazio, D. Hakim, *Morrey Spaces: Introduction and Applications to Integral Operators and PDE's*, vol. II (Taylor and Francis, New York, 2020)
19. S. Semmes, Metric spaces and mappings seen at many scales, in *Metric Structures for Riemannian and Non-Riemannian Spaces*, ed. by M. Gromov. Progress in Mathematics, vol. 152 (Birkhaüser, Boston, 1999)
20. N.S. Trudinger, On Harnack type inequalities and their application to quasilinear elliptic equations. CPAM **XX**, 721–747 (1967)
21. C. Vitanza, P. Zamboni, Necessary and sufficient conditions for Hölder continuity of solutions of degenerate Schrödinger operators. Le Matematiche **52**(2), 393–409 (1997)
22. P. Zamboni, Hölder continuity for solutions of linear degenerate elliptic equations under minimal assumptions. J. Differential Equations **182**(1), 121–140 (2002)

Monotonicity of Positive Solutions to $-\Delta_p u + a(u)|\nabla u|^q = f(u)$ in the Half-Plane in the Case $p \leqslant 2$

Luigi Montoro

Dedicated to Professor Emmanuele DiBenedetto in occasion of his 70th birthday.

Abstract We consider a quasilinear elliptic equation involving a first order term, under zero Dirichlet boundary condition in the half-plane. We prove that any positive possibly unbounded solution, is monotone increasing with respect to the direction orthogonal to the boundary.

Keywords p-Laplace operator · First order term · Half-plane · Qualitative properties

1 Introduction

We consider $C^{1,\alpha}$ weak solutions to the problem

$$\begin{cases} -\Delta_p u + a(u)|\nabla u|^q = f(u) & \text{in } \mathbb{R}^2_+ := \{(x, y) \in \mathbb{R}^2 \,:\, y > 0\} \\ u(x', y) > 0 & \text{in } \mathbb{R}^2_+ \\ u(x', 0) = 0 & \text{on } \partial\mathbb{R}^2_+, \end{cases} \tag{$\mathcal{P}$}$$

The author was partially supported by the Gnampa Project 2016 'Proprietà qualitative di equazioni ellittiche e paraboliche non lineari'.

L. Montoro (✉)
Dipartimento di Matematica e Informatica, Università della Calabria, Arcavacata di Rende, Cosenza, Italy
e-mail: montoro@mat.unical.it

V. Vespri et al. (eds.), *Harnack Inequalities and Nonlinear Operators*,
Springer INdAM Series 46, https://doi.org/10.1007/978-3-030-73778-8_8

where, $-\Delta_p u := -\operatorname{div}(|\nabla u|^{p-2}\nabla u)$ denotes the p-Laplace operator. Moreover, through all the paper, we assume that the following hypotheses (denoted by (hp^*) in the sequel) hold:

(hp^*) *(i)* $\frac{3}{2} < p \leqslant 2, 1 < q \leq p$;
(ii) a is a locally Lipschitz continuous function on $\mathbb{R}$ with Lipschitz constant denoted by L_a;
(iii) f is a locally Lipschitz continuous function on $\mathbb{R}_0^+$ with Lipschitz constant denoted by L_f and we assume that $f(s) > 0$ for $s > 0$.

Note that the $C^{1,\alpha}$ regularity of the solutions to problems involving the p-Laplace operator follows by the well known results in [18, 33, 34, 45]. To be more precise we assume that

$$\text{Given a compact set } \mathcal{K} \subset \mathbb{R}^2, \text{ we have that}$$

$$u \in C^{1,\alpha}(\mathcal{K} \cap \overline{\mathbb{R}^2_+}).$$

We study monotonicity properties of the solutions, w.r.t. (with respect to) the y-direction, via the Alexandrov-Serrin moving plane method [1, 3, 30, 44]. For the semilinear case, the founding papers on this topic go back to [4–7, 15, 29]. We also refer the readers to [8, 11, 16, 17, 20, 21, 23, 43] for other results concerning the monotonicity of the solutions in half-spaces also in more general settings (always in the uniformly elliptic case) and to [2, 9, 10, 19, 28, 31, 32, 41] (and the references therein) for some monotonicity and symmetry results in the nonlocal setting.

Here we consider the quasilinear problem ($\mathcal{P}$) and we continue the study that we have started in [24–27, 38, 40] (see also [22]): without assuming any hypothesis on the boundedness of the solution u or of the gradient ∇u, using a geometric approach, we prove in the half-plane the following

Theorem 1.1 *Let $u \in C^{1,\alpha}_{loc}(\overline{\mathbb{R}^2_+})$ be a weak solution to ($\mathcal{P}$) and let (hp^*) be fulfilled. Then u is monotone increasing w.r.t. the y-direction, that is*

$$\frac{\partial u}{\partial y} > 0 \quad in \quad \mathbb{R}^2_+.$$

We remark that, already in the case $p = 2$ and $a(u) = 0$ in ($\mathcal{P}$), there exist unbounded monotone solutions u whose gradient $|\nabla u|$ is also unbounded (think for example to $u(x, y) = e^x y$). This motivates our analysis.

When $a(u) = 0$ problem ($\mathcal{P}$) reduces to

$$\begin{cases} -\Delta_p u = f(u), & \text{in } \mathbb{R}^2_+ \\ u(x', y) > 0, & \text{in } \mathbb{R}^2_+ \\ u(x', 0) = 0, & \text{on } \partial\mathbb{R}^2_+ . \end{cases} \tag{1.1}$$

The monotonicity of solutions to (1.1) was first studied in [14] considering positive nonlinearities: the result holds under the restriction $p > 3/2$, that is the assumption that we require in the hypothesis (hp^*).

2 Preliminaries

In this section we recall some results related to the p-Laplace equations. First of all we point out the following inequalities that we will use in the sequel:

$\forall \eta, \eta' \in \mathbb{R}^N$ with $|\eta| + |\eta'| > 0$ there exist positive constants $\dot{C}, \check{C}$ depending on p such that

$$[|\eta|^{p-2}\eta - |\eta'|^{p-2}\eta'][\eta - \eta'] \geq \dot{C}(|\eta| + |\eta'|)^{p-2}|\eta - \eta'|^2, \tag{2.1}$$

$$||\eta|^{p-2}\eta - |\eta'|^{p-2}\eta'| \leqslant \check{C}(|\eta| + |\eta'|)^{p-2}|\eta - \eta'|.$$

Notation Generic fixed numerical constants will be denoted by C (with subscript in some case) and they will be allowed to vary within a single line or formula. $|S|$ denotes the Lebesgue measure of a set S. f^+ and f^- are the positive part and the negative part of a function f, i.e. $f^+ = \max\{f, 0\}$ and $f^- = -\min\{f, 0\}$.

Here below we recall some known and some new results related to the p-Laplace equations. In particular we refer to maximum and comparison principles. Recently in [36], it was obtained a Harnack comparison type inequality for C^1 weak solutions to general quasilinear problems under suitable general assumptions; in particular the results apply to $(\mathcal{P})$. An important consequence of the Harnack comparison inequality is in fact a strong comparison principle (for the general case we refer to [36, 37], see also [12, 13, 35, 39]) that we point out for our interest in the following

Theorem 2.1 (Strong Comparison and Maximum Principle) *Let $(2N+2)/(N+2) < p \leqslant 2$ and $u, v \in C^1(\bar{\Omega})$ with u a weak solution to*

$$-\Delta_p w + a(w)|\nabla w|^q = f(w) \quad in \quad \Omega.$$

(*i*) *Assume that $q \geq 1$ and assume that $f(u), a(u)$ fulfill (hp^*). Then, if Ω is a connected domain and if*

$$-\Delta_p u + a(u)|\nabla u|^q - f(u) \leqslant -\Delta_p v + a(v)|\nabla v|^q - f(v), \quad u \leqslant v \quad in \quad \Omega$$

in the weak distributional meaning, it follows that

$$u < v \quad in \quad \Omega$$

unless $u \equiv v$ in Ω.

(*ii*) *For any* $i \in \{1, 2, \ldots, N\}$ *and any connected domain* $\Omega' \subset \Omega$, *with* $u_{x_i} \geq 0$ *in* Ω', *we have* $u_{x_i} > 0$ *in* Ω', *unless* $u_{x_i} \equiv 0$ *in* Ω'.

We prove here a weak comparison principle in small domains that we need later in the proof of our main result. We have

Proposition 2.2 (Weak Comparison Principle in Small Domains) *Assume that the hypotheses* (hp^*) *hold and let* $u, v \in C^1(\overline{\Omega})$ *such that*

$$\begin{cases} -\Delta_p u + a(u)|\nabla u|^q \leqslant f(u), & \text{in } \Omega \\ -\Delta_p v + a(v)|\nabla v|^q \geq f(v), & \text{in } \Omega \\ u > 0, v > 0 & \text{in } \Omega \\ u \leqslant v, & \text{on } \partial\tilde{\Omega}, \end{cases} \tag{2.2}$$

where $\tilde{\Omega}$ *is a bounded set such that* $\tilde{\Omega} \subset \Omega$. *Then there exists a positive constant*

$$\delta = \delta(p, q, L_a, L_f, \|u\|_{L^\infty(\Omega)}, \|\nabla u\|_{L^\infty(\Omega)}, \|\nabla v\|_{L^\infty(\Omega)})$$

such that if we assume $|\tilde{\Omega}| \leqslant \delta$, *then it holds*

$$u \leqslant v \quad \text{in } \tilde{\Omega}.$$

Proof We define $w = (u - v)^+$. Clearly $w \in W_0^{1,p}(\tilde{\Omega})$ and thus we use it as test function in both equations in (2.2). Subtracting (and using (2.1)) we have

$$\begin{aligned} &\dot{C} \int_{\tilde{\Omega}} (|\nabla u| + |\nabla v|)^{p-2} |\nabla (u - v)^+|^2 \, dx \\ &\leqslant \left| \int_{\tilde{\Omega}} (a(u)|\nabla u|^q - a(v)|\nabla v|^q)(u - v)^+ \, dx \right| + \int_{\tilde{\Omega}} |f(u) - f(v)|(u - v)^+ \, dx. \end{aligned} \tag{2.3}$$

We arrange (2.3) as

$$\begin{aligned} &\dot{C} \int_{\tilde{\Omega}} (|\nabla u| + |\nabla v|)^{p-2} |\nabla (u - v)^+|^2 \, dx \\ &\leqslant \int_{\tilde{\Omega}} |a(u)| \Big| |\nabla u|^q - |\nabla v|^q \Big| (u - v)^+ \, dx + \int_{\tilde{\Omega}} |\nabla v|^q \Big| a(u) - a(v) \Big| (u - v)^+ \, dx \\ &+ \int_{\tilde{\Omega}} |f(u) - f(v)|(u - v)^+ \, dx \end{aligned}$$

and using (hp^*)

$$\int_{\tilde{\Omega}} (|\nabla u| + |\nabla v|)^{p-2} |\nabla (u-v)^+|^2 \, dx$$
$$\leqslant C \int_{\tilde{\Omega}} \Big| |\nabla u|^q - |\nabla v|^q \Big| (u-v)^+ \, dx + C \int_{\tilde{\Omega}} [(u-v)^+]^2 \, dx,$$

with $C = C(p, q, L_a, L_f, \|u\|_{L^\infty(\Omega)}, \|\nabla v\|_{L^\infty(\Omega)})$. Since $q > 1$, by mean value theorem we get

$$\int_{\tilde{\Omega}} (|\nabla u| + |\nabla v|)^{p-2} |\nabla (u-v)^+|^2 \, dx \tag{2.4}$$
$$\leqslant C \int_{\tilde{\Omega}} (|\nabla u| + |\nabla v|)^{q-1} |\nabla (u-v)^+| (u-v)^+ \, dx + C \int_{\tilde{\Omega}} [(u-v)^+]^2 \, dx,$$

and then

$$\int_{\tilde{\Omega}} (|\nabla u| + |\nabla v|)^{p-2} |\nabla (u-v)^+|^2 \, dx \tag{2.5}$$
$$\leqslant C \int_{\tilde{\Omega}} \frac{(|\nabla u| + |\nabla v|)^{q-1}}{(|\nabla u| + |\nabla v|)^{\frac{p-2}{2}}} (|\nabla u| + |\nabla v|)^{\frac{p-2}{2}} |\nabla (u-v)^+| (u-v)^+ \, dx$$
$$+ C \int_{\tilde{\Omega}} [(u-v)^+]^2 \, dx$$
$$\leqslant C \int_{\tilde{\Omega}} (|\nabla u| + |\nabla v|)^{\frac{p-2}{2}} |\nabla (u-v)^+| (u-v)^+ \, dx$$
$$+ C \int_{\tilde{\Omega}} [(u-v)^+]^2 \, dx,$$

with $C = C(p, q, L_a, L_f, \|u\|_{L^\infty(\Omega)}, \|\nabla u\|_{L^\infty(\Omega)}, \|\nabla v\|_{L^\infty(\Omega)})$ a positive constant. Applying Young inequality in the r.h.s of (2.5) we finally obtain

$$\int_{\tilde{\Omega}} (|\nabla u| + |\nabla v|)^{p-2} |\nabla (u-v)^+|^2 \, dx$$
$$\leqslant \varepsilon C \int_{\tilde{\Omega}} (|\nabla u| + |\nabla v|)^{p-2} |\nabla (u-v)^+|^2 \, dx + \frac{C}{\varepsilon} \int_{\tilde{\Omega}} [(u-v)^+]^2 \, dx$$
$$+ C \int_{\tilde{\Omega}} [(u-v)^+]^2 \, dx.$$

Finally for ε small we deduce

$$\int_{\tilde{\Omega}} (|\nabla u| + |\nabla v|)^{p-2} |\nabla (u-v)^+|^2 \, dx \leqslant C \int_{\tilde{\Omega}} [(u-v)^+]^2 \, dx \tag{2.6}$$

and $C = C(p, q, L_a, L_f, \|u\|_{L^\infty(\Omega)}, \|\nabla u\|_{L^\infty(\Omega)}, \|\nabla v\|_{L^\infty(\Omega)})$ a positive constant. The conclusion follows from (2.6) using Poincaré inequality: in fact the term $(|\nabla u| + |\nabla v|)^{p-2}$ is bounded away from zero since $|\nabla u|, |\nabla v| \in L^\infty(\Omega)$. Then (2.6) becomes

$$\int_{\tilde{\Omega}} |\nabla(u-v)^+|^2\, dx \leqslant C C_p(\tilde{\Omega}) \int_{\tilde{\Omega}} |\nabla(u-v)^+|^2\, dx,$$

with $C_p(\tilde{\Omega})$ denoting the standard Poincaré constant.

If $|\tilde{\Omega}| \leqslant \delta$, with $\delta = \delta(p, q, L_a, L_f, \|u\|_{L^\infty(\Omega)}, \|\nabla u\|_{L^\infty(\Omega)}, \|\nabla v\|_{L^\infty(\Omega)})$ sufficiently small, we may assume $C_p(\tilde{\Omega})$ such that

$$C C_p(\tilde{\Omega}) < 1.$$

This shows that actually $(u-v)^+ = 0$ in $\tilde{\Omega}$, concluding the proof. □

3 Proof of Theorem 1.1

Working in $\mathbb{R}^2$, the use of weak comparison principles in narrow unbounded domains is avoided by a geometrical argument that allows to use only a weak comparison principle in small domains. The main advantage is that we do not assume that either the solution u or gradient $|\nabla u|$ are bounded, a usual hypothesis in the literature in the case $N > 2$.

Before starting with the proof, let us introduce some necessary notation. Let $L_{x_0,s,\theta}$ be the line with slope $\tan(\theta)$ passing through the point (x_0, s), and V_θ the vector orthogonal to $L_{x_0,s,\theta}$ such that $\langle V_\theta, e_2\rangle \geqslant 0$. Denote by $\mathcal{T}_{x_0,s,\theta}$ the triangle delimited by $L_{x_0,s,\theta}$, $\{y = 0\}$ and $\{x = x_0\}$ (see Fig. 1). Define $T_{x_0,s,\theta}(x)$ as the point symmetric to x w.r.t. $L_{x_0,s,\theta}$ (see Fig. 2), and

$$u_{x_0,s,\theta}(x) = u(T_{x_0,s,\theta}(x)),$$

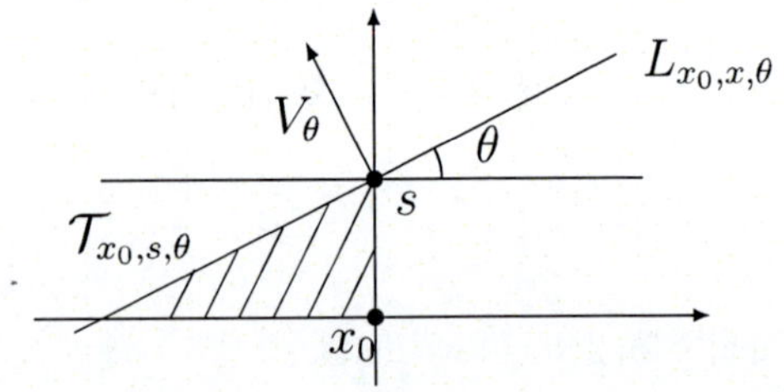

Fig. 1 The Triangle $\mathcal{T}_{x_0,s,\theta}$

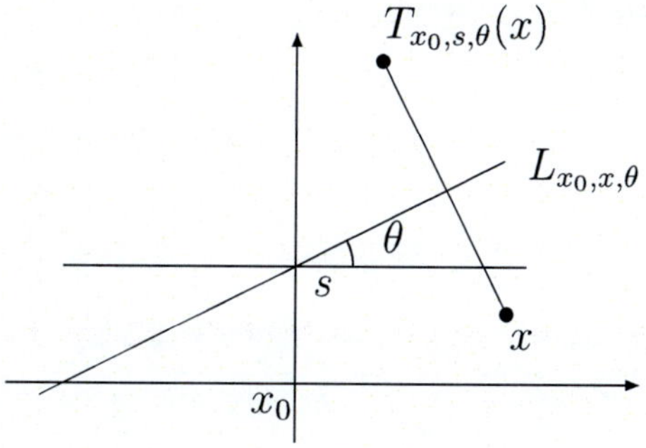

Fig. 2 Reflection w.r.t. $L_{x_0,s,\theta}$

and

$$w_{x_0,s,\theta} = u - u_{x_0,s,\theta}.$$

It is well known that $u_{x_0,s,\theta}$ still fulfills

$$-\Delta_p u_{x_0,s,\theta} + a(u_{x_0,s,\theta})|\nabla u_{x_0,s,\theta}|^q = f(u_{x_0,s,\theta}).$$

Moreover for simplicity we shall denote $u_{x_0,s,0} = u_s$.

Given any $x \in \mathbb{R}$, by Hopf boundary Lemma see [42, Theorem 5.5.1],

$$u_y(x,0) = \frac{\partial u}{\partial y}(x,0) > 0.$$

However, $u_y(x,0)$ possibly goes to 0 if $x \to \pm\infty$. So, we fix x_0 and h such that

$$\frac{\partial u}{\partial y}(x,y) \geqslant \gamma > 0, \quad \forall (x,y) \in Q_h(x_0),$$

where

$$Q_h(x_0) = \{(x,y) : \ |x - x_0| \leqslant h, \ 0 \leqslant y \leqslant 2h\}, \tag{3.1}$$

as shown in Fig. 3. Note that such a $\gamma > 0$ exists since $u \in C^{1,\alpha}$. Also, since $u \in C^{1,\alpha}$, we may assume that there exists

$$\delta_1 = \delta_1(h, \gamma, x_0) > 0 \tag{3.2}$$

such that, if $|\theta| \leqslant \delta_1$ (and consequently $V_\theta \approx e_2$), we have

$$\frac{\partial u}{\partial V_\theta} \geqslant \frac{\gamma}{2} > 0, \quad \text{in } Q_h(x_0). \tag{3.3}$$

Fig. 3 The Square $Q_h(x_0)$

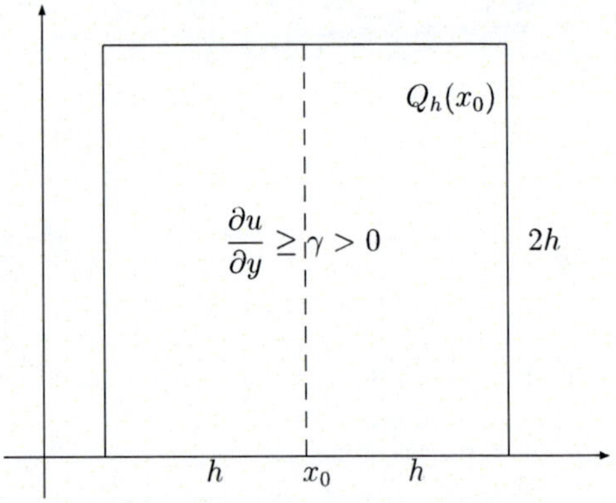

Fig. 4 The reflected Triangle

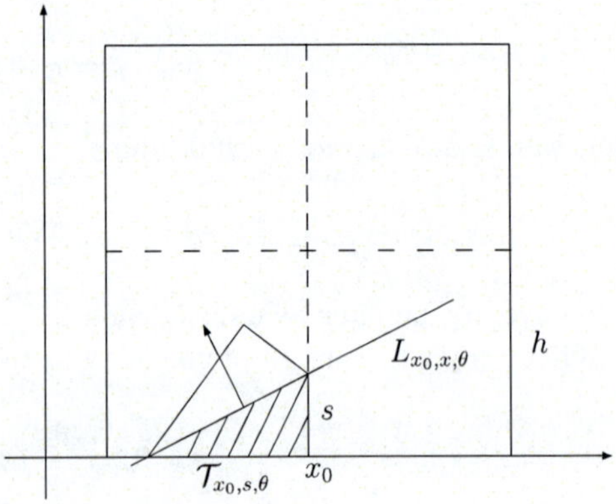

Claim 1 Let $Q_h(x_0)$ as in (3.1) and δ_1 defined in (3.2) and fix $\theta \neq 0$ with $|\theta| \leqslant \delta_1$. Then it is possible to find $\bar{s} = \bar{s}(\theta)$ such that for any $s \leqslant \bar{s}$ the triangle $\mathcal{T}_{x_0,s,\theta}$ is contained in $Q_h(x_0)$ and $u < u_{x_0,s,\theta}$ in $\mathcal{T}_{x_0,s,\theta}$ (with $u \leqslant u_{x_0,s,\theta}$ on $\partial\mathcal{T}_{x_0,s,\theta}$), see Fig. 4.

To prove Claim 1, fix θ such that $|\theta| \leqslant \delta_1$ and set $\bar{s} \leqslant h$ such that, for $s \leqslant \bar{s}$:

- The triangle $\mathcal{T}_{x_0,s,\theta}$ is contained in $Q_h(x_0)$ as well as the triangle obtained from $\mathcal{T}_{x_0,s,\theta}$ by reflection with respect to the line $L_{x_0,s,\theta}$ (see Fig. 4). Note that this is possible by simple geometric considerations.
- $u \leqslant u_{x_0,s,\theta}$ on $\partial\mathcal{T}_{x_0,s,\theta}$. In fact, since $|\theta| \leqslant \delta_1$ then $u \leqslant u_{x_0,s,\theta}$ on the line (x_0, y) for $0 \leqslant y \leqslant s$, as a consequence of the monotonicity in the V_θ-direction, by construction, see (3.3). Also $u \leqslant u_{x_0,s,\theta}$ if $y = 0$ by the Dirichlet assumption and the fact that u is positive in the interior of the domain. Finally $u \equiv u_{x_0,s,\theta}$ on $L_{x_0,s,\theta}$.

With this construction, for $0 < s \leqslant \bar{s}$, we have that

$$w_{x_0,s,\theta} = u - u_{x_0,s,\theta} \leqslant 0 \quad \text{on } \partial\mathcal{T}_{x_0,s,\theta}. \tag{3.4}$$

Possibly reducing $\bar{s}$, we can assume that the triangle $\mathcal{T}_{x_0,s,\theta}$ has sufficiently small measure in order to exploit the weak comparison principle in small domains, see Proposition 2.2. Then, from (3.4) (and Proposition 2.2) we get

$$w_{x_0,s,\theta} \leqslant 0 \quad \text{in } \mathcal{T}_{x_0,s,\theta}.$$

Also, since the case $w_{x_0,s,\theta} \equiv 0$ is clearly impossible, by the strong comparison principle (namely Theorem 2.1), we have

$$w_{x_0,s,\theta} < 0 \quad \text{in } \mathcal{T}_{x_0,s,\theta},$$

since the case $w_{x_0,s,\theta} \equiv 0$ is impossible. Therefore Claim 1 follows.

In the sequel we shall make repeated use of a technique which is the product of the *moving plane technique*, the *rotating plane technique* and the *sliding plane technique*. Let us explain next these techniques in an axiomatic way for future use.

Given (x_0, s, θ) and $\mathcal{T}_{x_0,s,\theta}$ as above, assume that

$$w_{x_0,s,\theta} \leqslant 0 \quad \text{on } \partial\mathcal{T}_{x_0,s,\theta}, \qquad \text{and} \qquad w_{x_0,s,\theta} < 0 \quad \text{in } \mathcal{T}_{x_0,s,\theta}, \tag{$\sharp$}$$

and suppose that for some (s', θ') sufficiently close to (s, θ) so that, $\mathcal{T}_{x_0,s',\theta'} \approx \mathcal{T}_{x_0,s,\theta}$, we have,

$$w_{x_0,s',\theta'} \leqslant 0 \quad \text{on } \partial\mathcal{T}_{x_0,s',\theta'}. \tag{$\flat$}$$

Since $w_{x_0,s,\theta} < 0$ in $\mathcal{T}_{x_0,s,\theta}$, we can fix a compact set $K \subset \mathcal{T}_{x_0,s,\theta}$ where $w_{x_0,s,\theta} \leqslant \rho < 0$. If (s', θ') are chosen appropriately close to (s, θ), we can assume without loss of generality that $K \subset \mathcal{T}_{x_0,s',\theta'}$,

$$w_{x_0,s',\theta'} \leqslant \frac{\rho}{2} < 0 \quad \text{in } K, \tag{3.5}$$

and the Lebesgue measure of $\mathcal{T}_{x_0,s',\theta'} \setminus K$ is small enough in order to apply the weak comparison principle (Proposition 2.2) in small domains.

Therefore, since $w_{x_0,s',\theta'} \leqslant 0$ on $\partial\big(\mathcal{T}_{x_0,s',\theta'} \setminus K\big)$ by ($\flat$) and (3.5), Proposition 2.2 yields,

$$w_{x_0,s',\theta'} \leqslant 0 \quad \text{in } \mathcal{T}_{x_0,s',\theta'} \setminus K$$

and consequently in the whole $\mathcal{T}_{x_0,s',\theta'}$. Then, by the strong comparison principle (Theorem 2.1), we get

$$w_{x_0,s',\theta'} < 0 \quad \text{in } \mathcal{T}_{x_0,s',\theta'}.$$

Summarizing, the outcome of the above argument is that after small translations and rotations, we can recover for $\mathcal{T}_{x0,s',\theta'}$ the same situation we initially had in $\mathcal{T}_{x0,s,\theta}$, that is ($\sharp$). More explicitly, we get that for (s', θ') sufficiently close to (s, θ),

$$w_{x0,s',\theta'} \leqslant 0 \quad \text{on } \partial\mathcal{T}_{x0,s',\theta'} \quad \text{and } w_{x0,s',\theta'} < 0 \quad \text{in } \mathcal{T}_{x0,s',\theta'}.$$

Let us now show that, the fact that we can make small translations and rotations of $\mathcal{T}_{x0,s,\theta}$ towards $\mathcal{T}_{x0,s',\theta'}$ when $(s', \theta') \approx (s, \theta)$, implies that we can also make larger translations and rotations.

More precisely, let us fix (s, θ) for which ($\sharp$) holds and let $(\bar{s}, \bar{\theta})$ be such that there exists a continuous function

$$\begin{aligned} g : [0, 1] &\to \mathbb{R}^2 \\ t &\mapsto (s(t), \theta(t)) \end{aligned}$$

with $g(0) = (s, \theta)$, $g(1) = (\bar{s}, \bar{\theta})$ and $\theta(t) \neq 0$ for every $t \in [0, 1)$. Finally, suppose that for every $t \in [0, 1)$

$$w_{x0,s(t),\theta(t)} \leqslant 0 \quad \text{on } \partial\mathcal{T}_{x0,s(t),\theta(t)} \quad \forall t \in [0, 1). \tag{3.6}$$

The above arguments imply that we can find some small $\tilde{t} > 0$ such that, for $0 < t \leqslant \tilde{t}$,

$$w_{x0,s(t),\theta(t)} \leqslant 0 \quad \text{on } \partial\mathcal{T}_{x0,s(t),\theta(t)} \quad \text{and} \quad w_{x0,s(t),\theta(t)} < 0 \quad \text{in } \mathcal{T}_{x0,s(t),\theta(t)}. \tag{3.7}$$

We now let,

$$\overline{T} \equiv \{\tilde{t} \in [0, 1] \ s.t. \text{(3.7) holds for any } 0 \leqslant t \leqslant \tilde{t}\} \quad \text{and set} \quad \bar{t} = \sup_{t \in \overline{T}} t.$$

Notice that we have already proved that $\bar{t} > 0$. The argument concludes by showing that, actually, $\bar{t} = 1$. To prove this, we proceed by contradiction and assume $\bar{t} < 1$. Then, by continuity

$$w_{x0,s(\bar{t}),\theta(\bar{t})} \leqslant 0 \quad \text{in } \mathcal{T}_{x0,s(\bar{t}),\theta(\bar{t})}$$

and, by the strong comparison principle

$$w_{x0,s(\bar{t}),\theta(\bar{t})} < 0 \quad \text{in } \mathcal{T}_{x0,s(\bar{t}),\theta(\bar{t})}.$$

We are now in the situation described in ($\sharp$) and ($\flat$). Therefore we can argue as above showing that it is still possible to push the plane slightly further, that is, finding a sufficiently small $\varepsilon > 0$ so that (3.7) holds for any $0 \leqslant t \leqslant \bar{t} + \varepsilon$, a contradiction with the definition of $\bar{t}$. This implies $\bar{t} = 1$. Summarizing, by means of this argument we get,

$$w_{x_0,\bar{s},\bar{\theta}} \leqslant 0 \quad \text{on } \partial\mathcal{T}_{x_0,\bar{s},\bar{\theta}} \quad \text{and} \quad w_{x_0,\bar{s},\bar{\theta}} < 0 \text{ in } \mathcal{T}_{x_0,\bar{s},\bar{\theta}}.$$

Now, we are going to apply the techniques just described. Let x_0, $Q_h(x_0)$ and δ_1 as in (3.1), (3.2). Define,

$$\Sigma_t = \{(x, y) \mid 0 < y < t\}.$$

We want to prove the following

Claim 2 Given any $\tilde{s}$ with $0 < \tilde{s} \leqslant h$, we have $u < u_{\tilde{s}}$ in $\Sigma_{\tilde{s}}$, and also clearly $u \leqslant u_{\tilde{s}}$ on $\partial\Sigma_{\tilde{s}}$ (recall that $u_{\tilde{s}}$ stands for $u_{x_0,\tilde{s},0}$).

To prove this, let us first fix θ such that $|\theta| \leqslant \delta_1$. Consequently, by Claim 1, we can find some $s = s(\theta) \leqslant \tilde{s}$ such that the triangle $\mathcal{T}_{x_0,s,\theta}$ is contained in $Q_h(x_0)$ (see Fig. 4), $u < u_{x_0,s,\theta}$ in $\mathcal{T}_{x_0,s,\theta}$ (and $u \leqslant u_{x_0,s,\theta}$ on $\partial\mathcal{T}_{x_0,s,\theta}$). Our purpose now is to enlarge the triangle $\mathcal{T}_{x_0,s,\theta}$ by applying the above arguments to the particular case of the transformation $g(t)$, namely, translations and rotations. The idea is to show that we can actually reach $\Sigma_{\tilde{s}}$ with these small perturbations of the initial triangle. In order to be able to do this, we shall have to check the hypothesis corresponding to ($\star$), ($\flat$) and (3.6).

Sliding Technique We start moving the line $L_{x_0,s,\theta}$ in the e_2-direction towards the line $L_{x_0,\tilde{s},\theta}$, keeping θ fixed and moving $s \to \tilde{s}$. In the notation above, we have $g(t) = (s(t), \theta)$ with $s(0) = s$ and $s(1) = \tilde{s}$, where in particular we can assume that $s(t) \leqslant \tilde{s}$. We note that for every $s(t) \leqslant \tilde{s}$ we have $u \leqslant u_{x_0,s(t),\theta}$ on $\partial\mathcal{T}_{x_0,s(t),\theta}$. To see this, notice that since $|\theta| \leqslant \delta_1$, then $u \leqslant u_{x_0,s(t),\theta}$ on the line (x_0, y) for $0 \leqslant y \leqslant s(t)$, because of the monotonicity in the V_θ-direction, that we have by construction. Also $u < u_{x_0,s(t),\theta}$ if $y = 0$ by the Dirichlet assumption, and the fact that u is positive in the interior of the domain. And finally, $u \equiv u_{x_0,s(t),\theta}$ on $L_{x_0,s(t),\theta}$ by definition. This shows that we have the right conditions ($\flat$) on the boundary for every $s(t) \leqslant \tilde{s}$; therefore by the technique described above, we get,

$$u < u_{x_0,\tilde{s},\theta} \text{ in } \mathcal{T}_{x_0,\tilde{s},\theta} \text{ and } u \leqslant u_{x_0,\tilde{s},\theta} \text{ on } \partial\mathcal{T}_{x_0,\tilde{s},\theta}.$$

Rotating Technique After having reached $\tilde{s}$, we start rotating the line $L_{x_0,\tilde{s},\theta}$ towards the line $\{y = \tilde{s}\}$, keeping $\tilde{s}$ fixed and letting $\theta \to 0$. That is, in the notation above, we consider $g(t) = (\tilde{s}, \theta(t))$ with $\theta(0) = \theta$ and $\theta(1) = 0$, where in particular we can assume that $\theta(t) \neq 0$ if $t \neq 1$. It can be easily checked, exactly in the same way as in the sliding technique, that we still have the right conditions on the boundary.

Note that to start from θ we need $|\theta| \leqslant \delta_1$. Therefore we can assume either that $0 < \theta(t) \leqslant \delta_1$, or that $0 < -\theta(t) \leqslant \delta_1$. Consider first the case when θ is positive. By the rotating plane technique, at the limit it follows that,

$$u(x, y) < u_{\tilde{s}}(x, y) = u(x, 2\tilde{s} - y) \qquad \text{in } \Sigma_{\tilde{s}} \cap \{x \leqslant x_0\}.$$

In the second case, we start from a negative θ, whence it follows that

$$u(x, y) < u_{\tilde{s}}(x, y) = u(x, 2\tilde{s} - y) \qquad \text{in } \Sigma_{\tilde{s}} \cap \{x \geqslant x_0\}.$$

Finally,

$$u(x, y) < u_{\tilde{s}}(x, y) = u(x, 2\tilde{s} - y) \qquad \text{in } \Sigma_{\tilde{s}}$$

for every $0 < \tilde{s} \leqslant h$, proving Claim 2.

We now point out some consequences. First, note that u is strictly monotone increasing in the e_2-direction in Σ_h. In fact, given (x, y_1) and (x, y_2) in Σ_h (say $0 \leqslant y_1 < y_2 \leqslant h$), we proved in Claim 2 that $u(x, y_1) < u_{\frac{y_1+y_2}{2}}(x, y_1)$ which yields,

$$u(x, y_1) < u(x, y_2).$$

This immediately gives $\partial u/\partial y \geqslant 0$ in Σ_h, but actually by Theorem 2.1 it follows

$$\partial u_y := \frac{\partial u}{\partial y} > 0 \quad \text{in } \Sigma_h. \tag{3.8}$$

Let us define,

$$\Lambda = \{\lambda \in \mathbb{R}^+ : u < u_{\lambda'} \text{ in } \Sigma_{\lambda'} \text{ for every } \lambda' < \lambda\}$$

and

$$\bar{\lambda} = \sup_{\lambda \in \Lambda} \lambda, \tag{3.9}$$

being Λ not empty, since from Claim 2, we know that $\bar{\lambda} \geq h > 0$. As before, $u \leqslant u_{\bar{\lambda}}$ by continuity, which implies $u < u_{\bar{\lambda}}$ by the strong comparison principle. Moreover, as in (3.8), we have

$$\frac{\partial u}{\partial y} > 0 \quad \text{in } \Sigma_{\bar{\lambda}}. \tag{3.10}$$

To finish the proof of theorem we have to show that actually

Claim 3 $\bar{\lambda} = \infty$.

The proof is by contradiction, so we assume that $\bar{\lambda} < \infty$. We shall show that we can find $\varepsilon > 0$ small enough so that,

$$u < u_{\lambda'} \quad \text{for } 0 < \lambda' \leqslant \bar{\lambda} + \varepsilon, \tag{3.11}$$

a contradiction with the definition of $\bar{\lambda}$. This would finish the proof of Theorem 1.1.

By continuity, from (3.10) (since $u \in C^1$), we infer that actually $\partial u_y(x, \bar{\lambda}) \geq 0$. We actually show that there exists some $x_0 \in \mathbb{R}$ such that

$$\frac{\partial u}{\partial y}(x_0, \bar{\lambda}) > 0. \tag{3.12}$$

To prove this we argue by contradiction and we assume that $\partial u_y(x\,, \bar{\lambda}) = 0$ for all $x \in \mathbb{R}$. Consider the function $u^\star(x, y)$ defined in $\Sigma_{2\bar{\lambda}}$ by

$$u^\star(x, y) = \begin{cases} u(x, y) & \text{if } \ 0 \leqslant y \leqslant \bar{\lambda} \\ u(x, 2\bar{\lambda} - y) & \text{if } \ \bar{\lambda} \leqslant y \leqslant 2\bar{\lambda} \end{cases}$$

and the function $u_\star(x, y)$ defined in $\Sigma_{2\bar{\lambda}}$ by

$$u_\star(x, y) = \begin{cases} u(x, 2\bar{\lambda} - y) & \text{if } \ 0 \leqslant y \leqslant \bar{\lambda} \\ u(x, y) & \text{if } \ \bar{\lambda} \leqslant y \leqslant 2\bar{\lambda}. \end{cases}$$

Note that $u^\star$ is the even reflection (w.r.t. the line $y = \bar{\lambda}$) of $u|_{\Sigma_{\bar{\lambda}}}$ and $u_\star$ is the even reflection (w.r.t. the line $y = \bar{\lambda}$) of $u|_{\Sigma_{2\bar{\lambda}} \setminus \Sigma_{\bar{\lambda}}}$. Since we are assuming that $\partial u_y(x\,, \bar{\lambda}) = 0$ for every $x \in \mathbb{R}$, it follows that $u^\star$ and $u_\star$ are C^1 solutions of $-\Delta_p w + a(w)|\nabla w|^q = f(w)$ in $\Sigma_{2\bar{\lambda}}$. Since we know that $u < u_{\bar{\lambda}}$ in $\Sigma_{\bar{\lambda}}$, we have $u^\star \leqslant u_\star$ in $\Sigma_{2\bar{\lambda}}$ and $u^\star$ does not coincide with $u_\star$ since of the strict inequality $u < u_{\bar{\lambda}}$ in $\Sigma_{\bar{\lambda}}$. However, since $u^\star(x, \bar{\lambda}) = u_\star(x, \bar{\lambda})$ for any $x \in \mathbb{R}$, by Theorem 2.1 it would follows that $u^\star \equiv u_\star$ in $\Sigma_{2\bar{\lambda}}$. This contradiction proves (3.12).

In the proof of Claim 3, we shall need the following result, whose proof follows verbatim from [6, 8, 14]; therefore we skip it.

Lemma 3.1 *Consider $x_0 \in \mathbb{R}$ and $\bar{\lambda} > 0$ such that,*

(1) $\frac{\partial u}{\partial y}(x_0\,, y) > 0$ *for every $y \in [0, \bar{\lambda}]$.*
(2) *For every $\lambda \in (0, \bar{\lambda}]$ we have $u(x_0, y) < u_\lambda(x_0, y) = u(x_0\,, 2\lambda - y)$ for $y \in [0\,, \lambda)$.*

Then there exists $\delta_2 > 0$ *such that, for any* θ *such that* $|\theta| \leqslant \delta_2$ *and for any* $\lambda \in (0, \bar{\lambda} + \delta_2]$, *we have:*

$$u(x_0, y) < u_{x_0,\lambda,\theta}(x_0, y) \quad \textit{for } y \in [0, \lambda).$$

We remark that thanks to Hopf Lemma, (3.10) and (3.12) we are in the hypothesis of Lemma 3.1.

Finally we are going to prove (3.11) with $\epsilon = \delta_2$ given by Lemma 3.1. Let us fix θ with $|\theta| \leqslant \delta_2$ and λ small enough so that Claim 1 applies. From Claim 1, we get that the triangle $\mathcal{T}_{x_0,\lambda,\theta}$ is contained in $Q_h(x_0)$ (see Fig. 4), and

$$u < u_{x_0,\lambda,\theta} \quad \text{in } \mathcal{T}_{x_0,\lambda,\theta} \quad \text{with } u \leqslant u_{x_0,\lambda,\theta} \quad \text{on } \partial\mathcal{T}_{x_0,\lambda,\theta}.$$

Following the proof in Claim 2 we now start sliding the line $L_{x_0,\lambda,\theta}$ in the e_2-direction towards the line $L_{x_0,\bar{\lambda}+\delta_2,\theta}$, keeping θ fixed and letting $\lambda \to \bar{\lambda} + \delta_2$. First we point out that the appropriate boundary conditions hold, namely for every $\lambda \leqslant \bar{\lambda} + \delta_2$ we have $u \leqslant u_{x_0,\lambda,\theta}$ on $\partial\mathcal{T}_{x_0,\lambda,\theta}$. In fact, since $|\theta| \leqslant \delta_2$ then by Lemma 3.1 $u < u_{x_0,\lambda,\theta}$ on the line (x_0, y) for $0 \leqslant y < \lambda$. As before, $u \leqslant u_{x_0,\lambda,\theta}$ if $y = 0$ by the Dirichlet assumption and finally $u = u_{x_0,\lambda,\theta}$ on $L_{x_0,\lambda,\theta}$. Therefore the sliding technique described above, yields,

$$u < u_{x_0,\bar{\lambda}+\delta_2,\theta} \text{ in } \mathcal{T}_{x_0,\bar{\lambda}+\delta_2,\theta}, \quad \text{and} \quad u \leqslant u_{x_0,\bar{\lambda}+\delta_2,\theta} \text{ on } \partial\mathcal{T}_{x_0,\bar{\lambda}+\delta_2,\theta}.$$

We would like to stress that in the application of the sliding and rotating techniques during the proof of Claim 2, it was crucial to ensure that the vertical side of the triangle $\mathcal{T}_{x_0,s,\theta}$ and the segment resulting from its reflection with respect to $L_{x_0,s,\theta}$ were always inside $Q_h(x_0)$, as this fact was necessary in order to check the right boundary conditions (♯) and (♭). Thanks to Lemma 3.1 we get that when the perturbations are small enough, the right conditions still hold even if the vertical side of the triangle is outside $Q_h(x_0)$ and we cannot rely on monotonicity anymore.

We now start rotating the line $L_{x_0,\bar{\lambda}+\delta_2,\theta}$ towards the line $y = \bar{\lambda} + \delta_2$, freezing $\bar{\lambda} + \delta_2$ and letting $\theta \to 0$ as in the rotating technique. Again, we use Lemma 3.1 to check that we have the right boundary conditions. Arguing as in Claim 2, if we keep θ positive then at the limit $\theta \to 0$ we get $u < u_{\bar{\lambda}+\delta_2}$ in $\Sigma_{\bar{\lambda}+\delta_2} \cap \{x \leqslant x_0\}$. Otherwise if θ is negative, it follows that $u < u_{\bar{\lambda}+\delta_2}$ in $\Sigma_{\bar{\lambda}+\delta_2} \cap \{x \geqslant x_0\}$. Finally,

$$u < u_{\bar{\lambda}+\delta_2} \quad \text{in } \Sigma_{\bar{\lambda}+\delta_2},$$

a contradiction with the definition of $\bar{\lambda}$. This proves Claim 3 and hence concludes the proof of Theorem 1.1.

References

1. A.D. Alexandrov, A characteristic property of the spheres. Ann. Mat. Pura Appl. **58**, 303–354 (1962)
2. B. Barrios, L. Montoro, B. Sciunzi, On the moving plane method for nonlocal problems in bounded domains. J. Anal. Math. **135**(1), 37–57 (2018)
3. H. Berestycki, L. Nirenberg, On the method of moving planes and the sliding method. Bolletin Soc. Brasil. Mat. Nova. Ser. **22**(1), 1–37 (1991)
4. H. Berestycki, L.A. Caffarelli, L. Nirenberg, Symmetry for elliptic equations in the halfspace, in *Boundary Value Problems for PDEs and Applications*, Vol. dedicated to E. Magenes, J. L. Lions et al., ed. (Masson, Paris, 1993), pp. 27–42
5. H. Berestycki, L. Caffarelli, L. Nirenberg, Inequalities for second order elliptic equations with applications to unbounded domains. Duke Math. J. **81**(2), 467–494 (1996)
6. H. Berestycki, L. Caffarelli, L. Nirenberg, Further qualitative properties for elliptic equations in unbounded domains. Ann. Scuola Norm. Sup. Pisa Cl. Sci. (4) **25**(1–2), 69–94 (1997)
7. H. Berestycki, L. Caffarelli, L. Nirenberg, Monotonicity for elliptic equations in an unbounded Lipschitz domain. Comm. Pure Appl. Math. **50**, 1089–1111 (1997)
8. F. Charro, L. Montoro, B. Sciunzi, Monotonicity of solutions of Fully nonlinear uniformly elliptic equations in the half-plane. J. Diff. Equations **251**(6), 1562–1579 (2011)
9. W. Chen, C. Li, B. Ou, Qualitative properties of solutions for an integral equation. Discrete Contin. Dynam. Syst. **12**(2), 347–354 (2005)
10. W. Chen, C. Li, B. Ou, Classification of solutions for an integral equation. Commun. Pure Appl. Math. **59**, 330–343 (2006)
11. L. Damascelli, F. Gladiali, Some nonexistence results for positive solutions of elliptic equations in unbounded domains. Rev. Mat. Iberoamericana **20**(1), 67–86 (2004)
12. L. Damascelli, B. Sciunzi. Regularity, monotonicity and symmetry of positive solutions of m-Laplace equations. J. Differential Equations **206**(2), 483–515 (2004)
13. L. Damascelli, B. Sciunzi, Harnack inequalities, maximum and comparison principles, and regularity of positive solutions of m-Laplace equations. Calc. Var. Partial Differential Equations **25**(2), 139–159 (2006)
14. L. Damascelli, B. Sciunzi, Monotonicity of the solutions of some quasilinear elliptic equations in the half-plane, and applications. Diff. Integral Equations **23**(5–6), 419–434 (2010)
15. E.N. Dancer, Some notes on the method of moving planes. Bull. Australian Math. Soc. **46**(3), 425–434 (1992)
16. E.N. Dancer, Some remarks on half space problems. Discrete Contin. Dynam. Syst. A **25**(1), 83–88 (2009)
17. E.N. Dancer, Y. Du, M. Efendiev, Quasilinear elliptic equations on half- and quarter-spaces. Adv. Nonlinear Stud. **13**, 115–136 (2013)
18. E. Di Benedetto, $C^{1+\alpha}$ local regularity of weak solutions of degenerate elliptic equations. Nonlinear Anal. **7**(8), 827–850 (1983)
19. S. Dipierro, L. Montoro, I. Peral, B. Sciunzi, Qualitative properties of positive solutions to nonlocal critical problems involving the Hardy-Leray potential. Calc. Var. Partial Differential Equations 55(4), 29 (2016), Paper No. 99
20. Y. Du, Z. Guo, Symmetry for elliptic equations in a half-space without strong maximum principle. Proc. R. Soc. Edinburgh Sect. A **134**(2), 259–269 (2004)
21. A. Farina, Rigidity and one-dimensional symmetry for semilinear elliptic equations in the whole of $\mathbb{R}^N$ and in half spaces. Adv. Math. Sci. Appl. **13**(1), 65–82 (2003)
22. A. Farina, B. Sciunzi, Qualitative properties and classification of nonnegative solutions to $-\Delta u = f(u)$ in unbounded domains when $f(0) < 0$. Rev. Mat. Iberoamericana **32**(4), 1311–1330 (2016)
23. A. Farina, E. Valdinoci, Flattening results for Elliptic PDEs in unbounded domains with applications to overdetermined problems. Arch. Rational Mech. Anal **195**(3), 1025 (2010)

24. A. Farina, L. Montoro, B. Sciunzi, Monotonicity and one-dimensional symmetry for solutions of $-\Delta_p u = f(u)$ in half-spaces. Calc. Var. Partial Differential Equations **43**, 123–145 (2012)
25. A. Farina, L. Montoro, B. Sciunzi, Monotonicity of solutions of quasilinear degenerate elliptic equations in half-spaces. Math. Ann. **357**(3), 855–893 (2013)
26. A. Farina, L. Montoro, G. Riey, B. Sciunzi, Monotonicity of solutions to quasilinear problems with a first-order term in half-spaces. Ann. Inst. H. Poincaré Anal. Non Linéaire **32**(1), 1–22 (2015)
27. A. Farina, L. Montoro, B. Sciunzi, Monotonicity in half-spaces of positive solutions to $-\Delta_p u = f(u)$ in the case $p > 2$. Ann. Scuola Norm. Sup. Pisa Cl. Sci. (5) **17**, 1207–1229 (2017)
28. P. Felmer, Y. Wang, Radial symmetry of positive solutions to equations involving the fractional Laplacian. Commun. Contemp. Math. **16**(1), 1350023, 24 pp. (2014)
29. B. Gidas, J. Spruck, A priori bounds for positive solutions of nonlinear elliptic equations. Comm. Partial Differential Equations **6**, 883–901 (1981)
30. B. Gidas, W.M. Ni, L. Nirenberg, Symmetry and related properties via the maximum principle. Comm. Math. Phys. **68**(3), 209–243 (1979)
31. S. Jarohs, T. Weth, Asymptotic symmetry for a class of nonlinear fractional reaction-diffusion equations. Discrete Contin. Dyn. Syst. **34**(6), 2581–2615 (2014)
32. S. Jarohs, T. Weth, Symmetry via antisymmetric maximum principles in nonlocal problems of variable order. Ann. Mat. Pura Appl. (4) **195**(1), 273–291 (2016)
33. O.A. Ladyzhenskaya, N.N. Uraltseva, *Linear and Quasilinear Elliptic Equations* (Academic Press, New York, 1968)
34. G.M. Lieberman, Boundary regularity for solutions of degenerate elliptic equations. Nonlinear Anal. **12**(11), 1203–1219 (1988)
35. S. Merchán, L. Montoro, I. Peral, B. Sciunzi, Existence and qualitative properties of solutions to a quasilinear elliptic equation involving the Hardy-Leray potential. Ann. Inst. H. Poincaré Anal. Non Linéaire **31**(1), 1–22 (2014)
36. S. Merchán, L. Montoro, B. Sciunzi, On the Harnack inequality for quasilinear elliptic equations with a first order term. Proc. Roy. Soc. Edinburgh Sect. A **148**(5), 1075–1095 (2018)
37. L. Montoro, Harnack inequalities and qualitative properties for some quasilinear elliptic equations. NoDEA Nonlinear Differential Equations Appl. **26**(6), 33 (2019), Paper No. 45
38. L. Montoro, B. Sciunzi, Monotonicity of the solutions of quasilinear elliptic equations in the half-plane with a changing sign nonlinearity. Contemp. Math. **528**, 71–79 (2010)
39. L. Montoro, B. Sciunzi, M. Squassina, Asymptotic symmetry for a class of quasi-linear parabolic problems. Adv. Nonlinear Stud. **10**(4), 789–818 (2010)
40. L. Montoro, G. Riey, B. Sciunzi, Qualitative properties of positive solutions to systems of quasilinear elliptic equations. Adv. Differential Equations **20**(7–8), 717–740 (2015)
41. L. Montoro, F. Punzo, B. Sciunzi, Qualitative properties of singular solutions to nonlocal problems. Ann. Mat. Pura Appl. (4) **197**(3), 941–964 (2018)
42. P. Pucci, J. Serrin, *The Maximum Principle* (Birkhauser, Boston, 2007)
43. A. Quaas, B. Sirakov, Existence results for nonproper elliptic equations involving the Pucci operator. Comm. Partial Differential Equations **31**(7–9), 987–1003 (2006)
44. J. Serrin, A symmetry problem in potential theory. Arch. Rational Mech. Anal **43**(4), 304–318 (1971)
45. P. Tolksdorf, Regularity for a more general class of quasilinear elliptic equations. J. Differential Equations **51**(1), 126–150 (1984)

Geometric Tangential Analysis and Sharp Regularity for Degenerate pdes

Eduardo V. Teixeira and José Miguel Urbano

To Emmanuele DiBenedetto, with admiration and friendship, on the occasion of his 70th birthday.

Abstract We provide a broad overview on qualitative versus quantitative regularity estimates in the theory of degenerate parabolic pdes. The former relates to DiBenedetto's revolutionary method of intrinsic scaling, while the latter is achieved by means of what has been termed geometric tangential analysis. We discuss, in particular, sharp estimates for the parabolic $p-$Poisson equation, for the porous medium equation and for the doubly nonlinear equation.

Keywords Degenerate parabolic equations · Sharp regularity · Tangential analysis · Intrinsic scaling

1 Introduction

In the mid 1980s, Emmanuele DiBenedetto made a series of crucial contributions (see, e.g., [17–19, 22, 23]) to the understanding of the regularity properties of weak solutions of singular and degenerate parabolic equations. His *method of intrinsic scaling* (cf. [20, 26, 27, 38] for rather complete accounts) would become a landmark in regularity theory, to be used extensively in the next decades to treat a variety of pdes, the most celebrated being the $p-$Laplace equation and the porous medium equation. The main insight supporting the method is that each degenerate pde must

E. V. Teixeira
University of Central Florida, Orlando, FL, USA
e-mail: eduardo.teixeira@ucf.edu

J. M. Urbano (✉)
University of Coimbra, CMUC, Department of Mathematics, Coimbra, Portugal

Department of Mathematics, Universidade Federal da Paraíba, João Pessoa, PB, Brazil
e-mail: jmurb@mat.uc.pt

V. Vespri et al. (eds.), *Harnack Inequalities and Nonlinear Operators*,
Springer INdAM Series 46, https://doi.org/10.1007/978-3-030-73778-8_9

be analysed in its own geometric setting, in which space and time scale according to the nature of the degeneracy. The concrete implementation of this general principle is rather involved, requiring the use of fine analytic estimates and sophisticated iterative methods (see, among others, [21, 24, 25]). Although powerful and versatile, the method of intrinsic scaling delivers essentially qualitative results, placing the weak solutions of certain pdes in the right regularity class but not providing sharp quantitative information. For example, it tells us that solutions of the porous medium equation

$$u_t - \operatorname{div}\left(m|u|^{m-1}\nabla u\right) = 0$$

are locally of class $C^{0,\alpha}$, for a certain (small) Hölder exponent α but is tight-lipped on the best possible α.

While, for many purposes, qualitative estimates for a given model are enough, the investigation pertaining to sharp estimates in diffusive pdes is, by no means, a mere fanciful inquire. On the contrary, sharp estimates reveal important nuances of the problem and play a decisive role in a finer analysis of the model. As a way of example, they are decisive in the investigation of problems involving free boundaries. As a structural attribute of a given pde, obtaining optimal regularity estimates for a given diffusive model is often a challenging problem. In what follows, we will describe a successful geometric approach which often leads to sharp, or at least improved, estimates in Hölder spaces; the method is inspired by geometric insights related to the notion of *tangent pde models*.

In the sequel, we will describe in more detail a few seminal ideas which foster intuitive insights leading to a technical apparatus supporting the method. We will also exemplify the power of those ideas in some concrete, relevant problems, namely in obtaining sharp estimates for the parabolic $p-$Poisson equation, for the porous medium equation and for pdes involving doubly nonlinearities.

2 Geometric Tangential Analysis

The abstract concept of Tangent is rather classical and widely spread in the realm of mathematical sciences. It bears a notion of approximation, usually involving more regular objects, from which one can infer pertinent information about the original entity. Probably one of the most well known examples of Tangent comes from the idea of differentiation, where one locally approximates a nonlinear map by a linear one. The acclaimed Inverse Function Theorem from Calculus asserts that if the linear approximation of a function f at a point a is injective and surjective, then so is f in a neighbourhood of a. This is a classical example where qualitative information on the approximating object is transferred to the approximated one.

While tangent lines to the graph of a function, or even tangent hyperplanes to manifolds are, per se, rather concrete manifestations of Tangent, this powerful mathematical concept transcendes to more abstract settings, ultimately yielding decisive breakthroughs.

In the lines of the analogy above, Geometric Tangential Analysis (GTA) refers to a constructive systematic approach based on the concept that a problem which enjoys greater regularity can be tangentially accessed by certain classes of pdes. By means of iterative arguments, the method then imports this regularity, properly corrected through the path used to access the tangential equation, to the original class. The roots of this idea likely go back to the foundation of De Giorgi's geometric measure theory of minimal surfaces, and accordingly, it is present in the development of the modern theory of free boundary problems.

Indeed, an instrumental argument in De Giorgi's geometric measure treatment of minimal surfaces is the so called *flatness improvement*. Roughly speaking, De Giorgi in [16] shows that if a minimal surface S is flat enough, say in B_1, then it is even flatter in $B_{1/2}$. At least as important as the theorem itself is the reasoning of its proof, which heuristically goes as follows: arguing by contradiction, one would produce a sequence of minimal surfaces S_j in B_1, that are $1/j$ flat with respect to a direction ν_j; however the aimed flatness improvement is not verified in $B_{1/2}$. By compactness arguments, an appropriate scaling of S_j converges to the graph of a function f, which ought to solve the linearised equation, namely $\Delta f = 0$. Since the limiting function f is very smooth, flatness improvement is verified for f. Thus, for j_0 sufficiently large, one reaches a contraction on the assumption that no flatness improvement was possible for S_{j_0}.

Such a revolutionary, seminal idea borne fruit in many other fields of research. In particular the motto *flatness implies regularity*, largely promoted by Caffarelli and collaborators, thrived in the theory of free boundary problems from the 1970s and 1980s (see, among others, [1, 2, 9–13]). Powerful methods and geometric insights designed for the study of free boundary problems evolved and, in the 1990s, played a decisive role in Caffarelli's work on fully non-linear elliptic pdes (cf. [14]) and, subsequently, in his studies on Monge-Ampère equations (see [15]). As for second order fully nonlinear elliptic equations, Caffarelli uses Krylov-Safonov Harnack inequality, designed for viscosity solutions, as a universal compactness device. He measures *closeness* between variable coefficient equations and constant coefficient equations by means of coefficient oscillation; no linearisation takes place. Ultimately, he shows that if the constant coefficient equation $F(x_0, D^2u) = 0$ has a good regularity theory, then $F(x, D^2u) = 0$ inherits some universal estimates, provided the coefficient oscillation is small enough.

Restricted to diffusive processes, perhaps a didactical way to contemplate GTA is by drawing connected dots. Each dot represents a *class* of elliptic or parabolic pdes and each path is a compactness theorem.

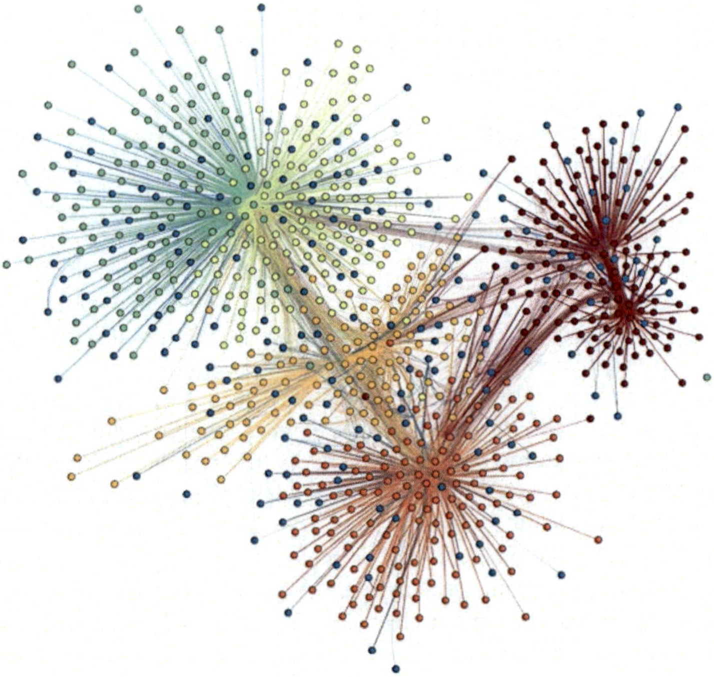

For instance, the study of all Poisson equations of the form $-\Delta u = f(x)$, say with $f \in L^p$, is one single dot. Caccioppoli-type energy estimates yield a connecting path from such dot to the (sought-after) dot representing all harmonic functions. Namely, if u_n is a sequence of functions, say bounded in L^2, satisfying $-\Delta u_n = f_n(x)$, if $f_n \to 0$ in L^p, then, up to a subsequence, $u_n \to h$ and h is harmonic.

The abstract concept of Tangent postulates that two connected PDEs should share an underlying regularity theory. The *caliber* of the path determines how much of the regularity one can bring from one model to another. In the above example, if

$$\Delta u = f(x), \quad f \in L^p,$$

then for any $0 < \lambda \ll 1$, the function $u_\lambda(x) := \frac{1}{\lambda^{2-\frac{n}{p}}} u(\lambda x)$ verifies

$$\Delta u_\lambda = f_\lambda(x),$$

where

$$f_\lambda(x) = \lambda^{\frac{n}{p}} f(\lambda x).$$

One easily verifies that

$$\|f_n(x)\|_p \leq \|f\|_p.$$

This means that through the path joining the Poisson equation, $-\Delta u = f(x) \in L^p$, and the Laplace equation, $\Delta h = 0$, one can transport estimates of order $O(r^{2-\frac{n}{p}})$. Such estimates ultimately yield $C^{0,2-\frac{n}{p}}$–regularity, if $0 < 2 - \frac{n}{p} < 1$, or $C^{1,1-\frac{n}{p}}$–regularity if $1 < 2 - \frac{n}{p} < 2$. As usual, the cases $p = n$ or $p = \infty$ are a bit tricker, as logarithm defects appear (see [34]).

In recent years, methods from Geometric Tangential Analysis have been significantly enhanced, amplifying their range of application and providing a more user-friendly platform for advancing these endeavours (cf. [5, 6, 31, 34–37], to cite just a few). In what follows, we shall present a small sample of problems that can be tackled by methods coming from GTA.

3 The Parabolic p–Poisson Equation

As a first example, we consider the degenerate parabolic p–Poisson equation

$$u_t - \operatorname{div}\left(|\nabla u|^{p-2}\nabla u\right) = f, \qquad p > 2, \tag{1}$$

with a source term $f \in L^{q,r}(U_T) \equiv L^r(0, T; L^q(U))$, where the exponents satisfy the conditions

$$\frac{1}{r} + \frac{n}{pq} < 1 \tag{2}$$

and

$$\frac{2}{r} + \frac{n}{q} > 1. \tag{3}$$

The first assumption is the standard minimal integrability condition that guarantees the existence of bounded weak solutions, while (3) defines the borderline setting for optimal Hölder type estimates. For instance, when $r = \infty$, conditions (2) and (3) enforce

$$\frac{n}{p} < q < n,$$

which corresponds to the known range of integrability required in the elliptic theory for local $C^{0,\alpha}$ estimates to be available.

We have shown in [37] that weak solutions are locally of class $C^{0,\alpha}$ in space, with

$$\alpha := \frac{(pq-n)r-pq}{q[(p-1)r-(p-2)]} = \frac{p\left(1-\frac{1}{r}-\frac{n}{pq}\right)}{\left(\frac{2}{r}+\frac{n}{q}-1\right)+p\left(1-\frac{1}{r}-\frac{n}{pq}\right)},$$

a precise and sharp expression for the Hölder exponent in terms of p, the integrability of the source and the space dimension n. Observe that $0 < \alpha < 1$, in view of (2) and (3).

We also have that u is of class $C^{0,\frac{\alpha}{\theta}}$ in time, where

$$\theta := \alpha + p - (p-1)\alpha = p - (p-2)\alpha = \alpha 2 + (1-\alpha)p$$

is the α–interpolation between 2 and p. If $p = 2$, we have $\theta = 2$. For $p > 2$, we have $2 < \theta < p$, since $0 < \alpha < 1$.

The regularity proof develops along the following lines.

Step 1: Closeness to p-caloric We first establish a key compactness result that states that if the source term f has a small norm in $L^{q,r}$, then a solution u to (1) is close to a p–caloric function in an inner subdomain. The proof is by contradiction and uses compactness driven from a Caccioppoli-type energy estimate and a control of the time derivative due to Lindqvist in [30].

Step 2: Geometric Iteration Then, we explore the approximation by p–caloric functions and the fact that p–caloric functions are *universally* Lipschitz continuous in space and $C^{0,\frac{1}{2}}$ in time. More precisely, we show there exist $\epsilon > 0$ and $0 < \lambda \ll 1/2$, depending only on p, n and α, such that if $\|f\|_{L^{q,r}(G_1)} \leq \epsilon$ and u is a local weak solution of (1) in G_1, with $\|u\|_{p,avg,G_1} \leq 1$, then there exists a convergent sequence of real numbers $\{c_k\}_{k\geq 1}$, with

$$|c_k - c_{k+1}| \leq c(n,p)\left(\lambda^{\alpha}\right)^k, \tag{4}$$

such that

$$\|u - c_k\|_{p,avg,G_{\lambda^k}} \leq \left(\lambda^k\right)^{\alpha}, \tag{5}$$

where the intrinsic θ–parabolic cylinder is defined by

$$G_\tau := \left(-\tau^\theta, 0\right) \times B_\tau(0), \quad \tau > 0$$

and the averaged norm is

$$\|v\|_{p,avg,Q} := \left(⨍_Q |v|^p \, dxdt\right)^{1/p} = |Q|^{-1/p}\|v\|_{p,Q}.$$

Step 3: Smallness Regime The smallness regime required is not restrictive since we can fall into that framework by scaling and contraction. Indeed, given a solution u, let

$$v(x,t) = \varrho u(\varrho^a x, \varrho^{(p-2)+ap} t)$$

(ϱ, a to be fixed), which is a solution of (1) with

$$\tilde{f}(x,t) = \varrho^{(p-1)+ap} f(\varrho^a x, \varrho^{(p-2)+ap} t).$$

Choose $a > 0$ such that $a < \frac{2}{n+p}$ and

$$[(p-1)+ap]r - a(n+p) - (p-2) > 0,$$

which is always possible (observe that the second condition holds for $a = 0$ and use its continuity with respect to a), and then $0 < \varrho < 1$ such that

$$\|v\|^p_{p,avg,G_1} \leq \varrho^{2-a(n+p)} \|u\|^p_{p,avg,G_1} \leq 1$$

and

$$\|\tilde{f}\|^r_{L^{q,r}(G_1)} = \varrho^{[(p-1)+ap]r - a(n+p) - (p-2)} \|f\|^r_{L^{q,r}(G_1)} \leq \epsilon^r.$$

Step 4: Hölder via Campanato Since the sequence $\{c_k\}$ is convergent, due to (4), let

$$\bar{c} := \lim_{k\to\infty} c_k.$$

It follows from (5) that, for arbitrary $0 < r < \frac{1}{2}$,

$$\fint_{G_r} |u - \bar{c}|^p \, dxdt \leq C r^{p\alpha}.$$

Standard covering arguments and the characterisation of Hölder continuity of Campanato–Da Prato give the local $C^{0;\alpha,\alpha/\theta}$—continuity.

To highlight the extent to which our result is sharp, we project it into the state of the art of the theory. For the linear case $p = 2$, we obtain

$$\alpha = 1 - \left(\frac{2}{r} + \frac{n}{q} - 1\right),$$

which is the optimal Hölder exponent for the non-homogeneous heat equation, and is in accordance with estimates obtained by energy considerations. When $p \to \infty$, we have $\alpha \to 1^-$, which gives an indication of the expected locally Lipschitz regularity for the case of the parabolic infinity-Laplacian. When the source f is independent of time, or else bounded in time, that is $r = \infty$, we obtain

$$\alpha = \frac{pq - n}{q(p-1)} = \frac{p}{p-1} \cdot \frac{q - \frac{n}{p}}{q},$$

which is exactly the optimal exponent obtained in [33] for the elliptic case.

Within the general theory of $p-$parabolic equations, our result reveals a surprising feature. From the applied point of view, it is relevant to know what is the effect on the diffusion properties of the model as we dim the exponent p. Naïve physical interpretations could indicate that the higher the value of p, the less efficient should the diffusion properties of the $p-$parabolic operator turn out to be, i.e., one should expect a less efficient smoothness effect of the operator. For instance, this is verified in the sharp regularity estimate for $p-$harmonic functions in the plane [29]. On the contrary, our estimate implies that for $p-$parabolic inhomogeneous equations, the Hölder regularity theory improves as p increases. In fact, a direct computation shows

$$\text{sign}\left(\partial_p \alpha(p, n, q, r)\right) = \text{sign}\left(q(2-r) + nr\right) = +1,$$

in view of standard assumptions on the integrability exponents of the source term.

4 The Porous Medium Equation

There are two crucial differences with respect to the previous case when treating the porous medium equation (cf. [39])

$$u_t - \text{div}\left(m|u|^{m-1}\nabla u\right) = f, \quad m > 1. \tag{6}$$

One is that adding a constant to a solution does not produce another solution, which somehow precludes the use of Campanato theory and requires a different technical approach to the Hölder regularity. The other is of a more fundamental nature, namely the hitherto unknown optimal regularity in the homogeneous case, leading to an extra dependence in the sharp Hölder exponent. Only for $n = 1$, it is proven in [7] that

$$\alpha_0 = \min\left\{1, \frac{1}{m-1}\right\}$$

but this is not the case in higher dimensions as corroborated by the celebrated counter-example in [8].

A locally bounded function

$$u \in C_{\rm loc}\left(0,T;L^2_{\rm loc}(U)\right), \qquad \text{with} \quad |u|^{\frac{m+1}{2}} \in L^2_{\rm loc}\left(0,T;W^{1,2}_{\rm loc}(U)\right)$$

is a local weak solution of (6) if, for every compact set $K \subset U$ and every subinterval $[t_1,t_2] \subset (0,T]$, we have

$$\int_K u\varphi\Big|_{t_1}^{t_2} + \int_{t_1}^{t_2}\int_K \left\{-u\varphi_t + m|u|^{m-1}\nabla u \cdot \nabla\varphi\right\} = \int_{t_1}^{t_2}\int_K f\varphi,$$

for all test functions

$$\varphi \in W^{1,2}_{\rm loc}\left(0,T;L^2(K)\right) \cap L^2_{\rm loc}\left(0,T;W^{1,2}_0(K)\right).$$

It is clear that all integrals in the above definition are convergent (cf. [26, §3.5]), interpreting the gradient term as

$$|u|^{m-1}\nabla u := \frac{2}{m+1}\,\mathrm{sign}(u)\,|u|^{\frac{m-1}{2}}\,\nabla|u|^{\frac{m+1}{2}}.$$

For a source term $f \in L^{q,r}(U_T) \equiv L^r(0,T;L^q(U))$, with

$$\frac{1}{r} + \frac{n}{2q} < 1,$$

it was shown in [4] that locally bounded weak solutions of (6) are locally of class $C^{0,\gamma}$ in space, with

$$\gamma = \frac{\alpha}{m}, \qquad \alpha = \min\left\{\alpha_0^-,\ \frac{m[(2q-n)r-2q]}{q[mr-(m-1)]}\right\}, \tag{7}$$

where $0 < \alpha_0 \leq 1$ denotes the optimal Hölder exponent for solutions of (6) with $f \equiv 0$. This regularity class is to be interpreted in the following sense: if

$$\frac{m[(2q-n)r-2q]}{q[mr-(m-1)]} < \alpha_0$$

then solutions are in $C^{0,\gamma}$, with

$$\gamma = \frac{(2q-n)r-2q}{q[mr-(m-1)]};$$

if, alternatively,

$$\frac{m[(2q-n)r-2q]}{q[mr-(m-1)]} \geq \alpha_0,$$

then solutions are in $C^{0,\gamma}$, for any $0 < \gamma < \frac{\alpha_0}{m}$.

Observe that

$$\frac{m[(2q-n)r-2q]}{q[mr-(m-1)]} = \frac{2m\left(1-\dfrac{1}{r}-\dfrac{n}{2q}\right)}{m\left(1-\dfrac{1}{r}\right)+\dfrac{1}{r}} > 0$$

and so indeed $\gamma > 0$. Note also that

$$\frac{m[(2q-n)r-2q]}{q[mr-(m-1)]} > 1$$

if

$$\left(1+\frac{1}{m}\right)\frac{1}{r}+\frac{n}{q} < 1,$$

and, as $q, r \to \infty$,

$$\frac{m[(2q-n)r-2q]}{q[mr-(m-1)]} \longrightarrow 2,$$

which means that after a certain integrability threshold it is the optimal regularity exponent of the homogeneous case that prevails, with

$$\alpha = \alpha_0^- \qquad \text{and} \qquad \gamma < \frac{\alpha_0}{m} < 1.$$

The $C^{0,\frac{\gamma}{\theta}}$ regularity in time is also obtained in [4], with

$$\theta = 2 - \left(1-\frac{1}{m}\right)\alpha = \alpha\left(1+\frac{1}{m}\right) + (1-\alpha)\,2$$

being the $\alpha-$interpolation between $1+\frac{1}{m}$ and 2. Observe that for $m=1$ we obtain

$$\gamma = 1 - \left(\frac{2}{r}+\frac{n}{q}-1\right) \qquad \text{and} \qquad \theta = 2,$$

recovering the optimal Hölder regularity for the non-homogeneous heat equation, in accordance with estimates obtained by energy considerations.

It is worth stressing that, as in the case of the p-Laplace equation, the integrability in time (respectively, in space) of the source affects the regularity in space (respectively, in time) of the solution. Here is a snapshot of the regularity proof.

Step 1: A Caccioppoli Estimate An equivalent definition of weak solution involving the Steklov average is instrumental in obtaining the following Caccioppoli estimate (cf. [26, §3.6]), for a constant C, depending only on n, m and $K \times [t_1, t_2]$:

$$\sup_{t_1<t<t_2} \int_K u^2\xi^2 + \int_{t_1}^{t_2} \int_K |u|^{m-1}|\nabla u|^2\xi^2$$

$$\leq C \int_{t_1}^{t_2} \int_K u^2\xi\, |\xi_t| + \int_{t_1}^{t_2} \int_K |u|^{m+1} \left(|\nabla\xi|^2 + \xi^2\right) + C\|f\|^2_{L^{q,r}},$$

for all $\xi \in C_0^\infty(K \times (t_1, t_2))$ such that $\xi \in [0, 1]$.

Step 2: The Intrinsic Geometric Setting It is crucial we perform our analysis in the adequate geometric setting, reflecting the degeneracy in the pde in the scaling of the space-time cylinders where the oscillation is measured. Given $0 < \alpha \leq 1$, let

$$\theta := 2 - \left(1 - \frac{1}{m}\right)\alpha \tag{8}$$

and define the intrinsic θ-parabolic cylinder as

$$G_\rho := \left(-\rho^\theta, 0\right) \times B_\rho(0), \quad \rho > 0.$$

Note that when $m = 1$, we obtain the standard parabolic cylinders reflecting the natural homogeneity between space and time for the heat equation. Observe also that θ satisfies the bounds

$$1 + \frac{1}{m} \leq \theta < 2,$$

which will be instrumental in the sequel.

Step 3: Link with the Homogeneous Case Another vital ingredient is a way to link solutions of the non-homogeneous problem with solutions of the homogeneous case. This type of statement follows from the available compactness for the problem and is proved by contradiction. Using the Caccioppoli estimate and the Arzelà-Ascoli theorem, we show that for every $\delta > 0$, there exists $0 < \epsilon \ll 1$ such that, if $\|f\|_{L^{q,r}(G_1)} \leq \epsilon$ and u is a local weak solution of (6) in G_1, with $\|u\|_{\infty,G_1} \leq 1$, then there exists ϕ such that

$$\phi_t - \operatorname{div}\left(m|\phi|^{m-1}\nabla\phi\right) = 0 \quad \text{in } G_{1/2}$$

and

$$\|u - \phi\|_{\infty, G_{1/2}} \leq \delta.$$

Step 4: Geometric Iteration This is the heart of the proof. We show that there exist $\epsilon > 0$ and $0 < \lambda \ll 1/2$, depending only on m, n and α, such that, if $\|f\|_{L^{q,r}(G_1)} \leq \epsilon$ and u is a local weak solution of (6) in G_1, with $\|u\|_{\infty, G_1} \leq 1$, then, for every $k \in \mathbb{N}$,

$$\|u\|_{\infty, G_{\lambda^k}} \leq (\lambda^k)^\gamma,$$

provided

$$|u(0,0)| \leq \frac{1}{4} \left(\lambda^k\right)^\gamma.$$

Recall that γ was fixed in (7) and let us proceed by induction. We start with the case $k = 1$. Take $0 < \delta < 1$, to be chosen later, and apply Step 3 to obtain $0 < \epsilon \ll 1$ and a solution ϕ of the homogeneous pme in $G_{1/2}$ such that

$$\|u - \phi\|_{\infty, G_{1/2}} \leq \delta.$$

Since ϕ is locally $C_x^{\alpha_0} \cap C_t^{\alpha_0/2}$, we obtain

$$\sup_{(x,t) \in G_\lambda} |\phi(x,t) - \phi(0,0)| \leq C \lambda^{\frac{\alpha_0}{m}},$$

for $C > 1$ universal, where $\lambda \ll 1$ is still to be chosen. In fact, for $(x,t) \in G_\lambda$,

$$\begin{aligned}
|\phi(x,t) - \phi(0,0)| &\leq |\phi(x,t) - \phi(0,t)| + |\phi(0,t) - \phi(0,0)| \\
&\leq c_1 |x - 0|^{\alpha_0} + c_2 |t - 0|^{\alpha_0/2} \\
&\leq c_1 \lambda^{\alpha_0} + c_2 \lambda^{\frac{\theta}{2}\alpha_0} \\
&\leq C \lambda^{\frac{\alpha_0}{m}},
\end{aligned}$$

since $\theta \geq 1 + \frac{1}{m} > \frac{2}{m}$. We can therefore estimate

$$\begin{aligned}
\sup_{G_\lambda} |u| &\leq \sup_{G_{1/2}} |u - \phi| + \sup_{G_\lambda} |\phi - \phi(0,0)| \\
&\quad + |\phi(0,0) - u(0,0)| + |u(0,0)| \\
&\leq 2\delta + C \lambda^{\frac{\alpha_0}{m}} + \frac{1}{4} \lambda^\gamma
\end{aligned}$$

and the result follows from the choices

$$\lambda = \left(\frac{1}{4C}\right)^{\frac{m}{\alpha_0-\alpha}} \qquad \text{and} \qquad \delta = \frac{1}{4}\lambda^{\gamma}.$$

Now suppose the conclusion holds for k and let's show it also holds for $k+1$. Due to (8), the function $v : G_1 \to \mathbb{R}$ defined by

$$v(x,t) = \frac{u(\lambda^k x, \lambda^{k\theta} t)}{\lambda^{\gamma k}}$$

solves

$$v_t - \operatorname{div}\left(m|v|^{m-1}\nabla v\right) = \lambda^{k(2-\alpha)} f(\lambda^k x, \lambda^{k\theta} t) = \tilde{f}(x,t).$$

We have

$$\begin{aligned}
\|\tilde{f}\|^r_{L^{q,r}(G_1)} &= \int_{-1}^{0} \left(\int_{B_1} \left|\tilde{f}(x,t)\right|^q\right)^{r/q} \\
&= \int_{-1}^{0} \left(\int_{B_1} \lambda^{k(2-\alpha)q} \left|f(\lambda^k x, \lambda^{k\theta} t)\right|^q\right)^{r/q} \\
&= \int_{-1}^{0} \left(\int_{B_{\lambda^k}} \lambda^{k(2-\alpha)q-kn} \left|f(x, \lambda^{k\theta} t)\right|^q\right)^{r/q} \\
&= \lambda^{[k(2-\alpha)q-kn]\frac{r}{q}} \int_{-1}^{0} \left(\int_{B_{\lambda^k}} \left|f(x, \lambda^{k\theta} t)\right|^q\right)^{r/q} \\
&= \lambda^{[k(2-\alpha)q-kn]\frac{r}{q}-k\theta} \int_{-\lambda^{k\theta}}^{0} \left(\int_{B_{\lambda^k}} |f(x,t)|^q\right)^{r/q}
\end{aligned}$$

and, since

$$[(2-\alpha)q - n]\frac{r}{q} - \theta \geq 0$$

due to (7), we get

$$\|\tilde{f}\|_{L^{q,r}(G_1)} \leq \|f\|_{L^{q,r}((-\lambda^{\theta k},0)\times B_{\lambda^k})} \leq \|f\|_{L^{q,r}(G_1)} \leq \epsilon,$$

which entitles v to the case $k = 1$. Note that $\|v\|_{\infty,G_1} \leq 1$, due to the induction hypothesis, and

$$|v(0,0)| = \left|\frac{u(0,0)}{\left(\lambda^k\right)^\gamma}\right| \leq \left|\frac{\frac{1}{4}\left(\lambda^{k+1}\right)^\gamma}{\left(\lambda^k\right)^\gamma}\right| \leq \frac{1}{4}\lambda^\gamma.$$

It then follows that

$$\|v\|_{\infty,G_\lambda} \leq \lambda^\gamma,$$

which is the same as

$$\|u\|_{\infty,G_{\lambda^{k+1}}} \leq \left(\lambda^{k+1}\right)^\gamma,$$

and the induction is complete.

Step 5: The Smallness Regime The final ingredient is showing that the smallness regime previously required is not restrictive. In fact, we can show that if u is a local weak solution of (6) in G_1 then, for every $0 < r < \lambda$,

$$\|u\|_{\infty,G_r} \leq C r^\gamma, \tag{9}$$

provided

$$|u(0,0)| \leq \frac{1}{4} r^\gamma. \tag{10}$$

To see this, take

$$v(x,t) = \rho u\left(\rho^a x, \rho^{(m-1)+2a} t\right)$$

with ρ, a to be fixed. It solves

$$v_t - \operatorname{div}(m|v|^{m-1}\nabla v) = \rho^{m+2a} f(\rho^a x, \rho^{(m-1)+2a)} t) = \tilde{f}(x,t)$$

and satisfies the bounds

$$\|v\|_{\infty,G_1} \leq \rho \|u\|_{\infty,G_1}$$

and

$$\|\tilde{f}\|^r_{L^{q,r}(G_1)} = \rho^{(m+2a)r - a(n\frac{r}{q}+2) - (m-1)} \|f\|^r_{L^{q,r}(G_1)}.$$

Choosing $a > 0$ such that

$$(m + 2a)r - a\left(\frac{nr}{q} + 2\right) - (m - 1) > 0,$$

which is always possible, and $0 < \rho \ll 1$, we enter the smallness regime. Now, given $0 < r < \lambda$, there exists $k \in \mathbb{N}$ such that

$$\lambda^{k+1} < r \leq \lambda^k.$$

Since $|u(0,0)| \leq \frac{1}{4}r^\gamma \leq \frac{1}{4}(\lambda^k)^\gamma$, it follows that

$$\|u\|_{\infty, G_{\lambda^k}} \leq (\lambda^k)^\gamma$$

and, for $C = \lambda^{-\gamma}$,

$$\|u\|_{\infty, G_r} \leq \|u\|_{\infty, G_{\lambda^k}} \leq (\lambda^k)^\gamma < \left(\frac{r}{\lambda}\right)^\gamma = C r^\gamma.$$

Step 6: Sharp Hölder Regularity We finally show there exists a universal constant K such that

$$\|u - u(0,0)\|_{\infty, G_r} \leq K r^\gamma,$$

which is the $C^{0,\gamma}$ regularity at the origin. We know, a priori, that u is continuous so

$$\mu := (4|u(0,0)|)^{1/\gamma} \geq 0$$

is well defined. Taking any radius $0 < r < \lambda$, we have three alternative cases.

- If $\mu \leq r < \lambda$, then (10) holds and

$$\sup_{G_r} |u(x,t) - u(0,0)| \leq C r^\gamma + |u(0,0)| \leq \left(C + \frac{1}{4}\right) r^\gamma$$

  follows from (9).
- If $0 < r < \mu$, we consider the function

$$w(x,t) := \frac{u(\mu x, \mu^\theta t)}{\mu^\gamma},$$

  which solves a uniformly parabolic equation in G_{ρ_0}, for a radius ρ_0 depending only on the data. This gives an estimate that, written in terms of u, reads

$$\sup_{G_r} |u(x,t) - u(0,0)| \leq C r^\gamma, \quad \forall\, 0 < r < \mu \frac{\rho_0}{2}.$$

- Finally, for $\mu\frac{\rho_0}{2} \leq r < \mu$, we have

$$\begin{aligned}\sup_{G_r} |u(x,t) - u(0,0)| &\leq \sup_{G_\mu} |u(x,t) - u(0,0)| \\ &\leq C\mu^\gamma \leq C\left(\frac{2r}{\rho_0}\right)^\gamma = \tilde{C}r^\gamma.\end{aligned}$$

5 The Doubly Nonlinear Equation

The methods and techniques of the previous two sections can be combined to obtain sharp regularity results for solutions of the inhomogeneous degenerate *doubly nonlinear parabolic equation*

$$u_t - \operatorname{div}\left(m|u|^{m-1}|\nabla u|^{p-2}\nabla u\right) = f, \qquad p > 2 \quad m > 1,$$

which appears, for example, in the contexts of non-Newtonian fluid dynamics, plasma physics, ground water problems or image processing.

The local Hölder continuity of bounded weak solutions is established in [28, 32]. For a source term $f \in L^{q,r}(U_T) \equiv L^r(0,T; L^q(U))$, with

$$\frac{1}{r} + \frac{n}{pq} < 1 \qquad \text{and} \qquad \frac{3}{r} + \frac{n}{q} > 2,$$

it is proven in [3] that locally bounded weak solutions are locally of class $C^{0,\beta}$ in space with

$$\beta = \frac{\alpha(p-1)}{m+p-2}, \qquad \alpha = \min\left\{\alpha_*^-, \frac{(m+p-2)[(pq-n)r-pq]}{q(p-1)[(r-1)(m+p-2)+1]}\right\},$$

where $0 < \alpha_* \leq 1$ denotes the optimal (unknown) Hölder exponent for solutions of the homogeneous case. The regularity class is to be interpreted as in the case of the porous medium equation.

Observe that when $m = 1$, the equation becomes the degenerate parabolic p−Poisson equation, for which $\alpha_* = 1$, and we recover the exponent

$$\alpha := \frac{(pq-n)r-pq}{q[(p-1)r-(p-2)]}$$

of Sect. 3. For $p = 2$, we have the porous medium equation and obtain the exponent (7) of Sect. 4.

Acknowledgments J.M.U. partially supported by the Center for Mathematics of the University of Coimbra, UID/MAT/00324/2013, funded by the Portuguese government through FCT/MCTES and co-funded by the European Regional Development Fund through Partnership Agreement PT2020.

References

1. H. Alt, L.A. Caffarelli, Existence and regularity for a minimum problem with free boundary. J. Reine Angew. Math. **325**, 105–144 (1981)
2. H. Alt, L.A. Caffarelli, A. Friedman, Variational problems with two phases and their free boundaries. Trans. Am. Math. Soc. **282**, 431–461 (1984)
3. J.G. Araújo, Sharp regularity for the degenerate doubly nonlinear parabolic equation. J. Differ. Equations **269**, 10558–10570 (2020)
4. D.J. Araújo, A. Maia, J.M. Urbano, Sharp regularity for the inhomogeneous porous medium equation. J. Anal. Math. **140**, 395–407 (2020)
5. D.J. Araújo, E.V. Teixeira, J.M. Urbano, A proof of the $C^{p'}$-regularity conjecture in the plane. Adv. Math. **316**, 541–553 (2017)
6. D.J. Araújo, E.V. Teixeira, J.M. Urbano, Towards the $C^{p'}$-regularity conjecture in higher dimensions. Int. Math. Res. Not. IMRN (2017). https://doi.org/10.1093/imrn/rnx068
7. D.G. Aronson, L.A. Caffarelli, Optimal regularity for one-dimensional porous medium flow. Rev. Mat. Iberoamericana **2**, 357–366 (1986)
8. D.G. Aronson, J. Graveleau, A self-similar solution to the focusing problem for the porous medium equation. European J. Appl. Math. **4**, 65–81 (1993)
9. L.A. Caffarelli, The regularity of free boundaries in higher dimensions. Acta Math. **139**, 155–184 (1977)
10. L.A. Caffarelli, Compactness methods in free boundary problems. Comm. Partial Differ. Equations **5**, 427–448 (1980)
11. L.A. Caffarelli, A Harnack inequality approach to the regularity of free boundaries. I. Lipschitz free boundaries are $C^{1,\alpha}$. Rev. Mat. Iberoamericana **3**, 139–162 (1987)
12. L.A. Caffarelli, A Harnack inequality approach to the regularity of free boundaries. III. Existence theory, compactness, and dependence on X. Ann. Scuola Norm. Sup. Pisa Cl. Sci. (4) **15**, 583–602 (1988)
13. L.A. Caffarelli, A Harnack inequality approach to the regularity of free boundaries. II. Flat free boundaries are Lipschitz. Comm. Pure Appl. Math. **42**, 55–78 (1989)
14. L.A. Caffarelli, Interior a priori estimates for solutions of fully nonlinear equations. Ann. Math. (2) **130**, 189–213 (1989)
15. L.A. Caffarelli, Interior $W^{2,p}$ estimates for solutions of the Monge-Ampère equation. Ann. Math. (2) **131**, 135–150 (1990)
16. E. De Giorgi, Frontiere orientate di misura minima, in *Seminario di Matematica della Scuola Normale Superiore di Pisa, 1960–1961* (Editrice Tecnico Scientifica, Pisa, 1961), 57 pp.
17. E. DiBenedetto, Continuity of weak solutions to certain singular parabolic equations. Ann. Mat. Pura Appl. (4) **130**, 131–176 (1982)
18. E. DiBenedetto, Continuity of weak solutions to a general porous medium equation. Indiana Univ. Math. J. **32**, 83–118 (1983)
19. E. DiBenedetto, $C^{1+\alpha}$ local regularity of weak solutions of degenerate elliptic equations. Nonlinear Anal. **7**, 827–850 (1983)
20. E. DiBenedetto, *Degenerate Parabolic Equations* (Universitext, Springer, New York, 1993)
21. E. DiBenedetto, Y.Z. Chen, Hölder estimates of solutions of singular parabolic equations with measurable coefficients. Arch. Ration. Mech. Anal. **118**, 257–271 (1992)
22. E. DiBenedetto, A. Friedman, Regularity of solutions of nonlinear degenerate parabolic systems. J. Reine Angew. Math. **349**, 83–128 (1984)

23. E. DiBenedetto, A. Friedman, Hölder estimates for nonlinear degenerate parabolic systems. J. Reine Angew. Math. **357**, 1–22 (1985)
24. E. DiBenedetto, U. Gianazza, V. Vespri, Subpotential lower bounds for nonnegative solutions to certain quasi-linear degenerate parabolic equations. Duke Math. J. **143**, 1–15 (2008)
25. E. DiBenedetto, U. Gianazza, V. Vespri, Harnack estimates for quasi-linear degenerate parabolic differential equations. Acta Math. **200**, 181–209 (2008)
26. E. DiBenedetto, U. Gianazza, V. Vespri, Harnack's inequality for degenerate and singular parabolic equations. *Springer Monographs in Mathematics* (Springer, New York, 2012)
27. E. DiBenedetto, J.M. Urbano, V. Vespri, Current issues on singular and degenerate evolution equations, in *Evolutionary Equations*. Handbook Differential Equation, vol. I (North-Holland, Amsterdam, 2004), pp. 169–286
28. A.V. Ivanov, Hölder estimates for quasilinear doubly degenerate parabolic equations. Zap. Nauchn. Sem. Leningrad. Otdel. Mat. Inst. Steklov. (LOMI) **171** (1989). Kraev. Zadachi Mat. Fiz. i Smezh. Voprosy Teor. Funktsiĭ **20**, 70–105, 185; translation in J. Soviet Math. **56**, 2320–2347 (1991)
29. T. Iwaniec, J.J. Manfredi, Regularity of p-harmonic functions on the plane. Rev. Mat. Iberoamericana **5**, 1–19 (1989)
30. P. Lindqvist, On the time derivative in a quasilinear equation. Trans. R. Norw. Soc. Sci. Lett. **2**, 1–7 (2008)
31. E. Pimentel, E.V. Teixeira, Sharp Hessian integrability estimates for nonlinear elliptic equations: an asymptotic approach. J. Math. Pures Appl. (9) **106**, 744–767 (2016)
32. M.M. Porzio, V. Vespri, Hölder estimates for local solutions of some doubly nonlinear degenerate parabolic equations. J. Differ. Equations **103**, 146–178 (1993)
33. E.V. Teixeira, Sharp regularity for general Poisson equations with borderline sources. J. Math. Pures Appl. (9) **99**, 150–164 (2013)
34. E.V. Teixeira, Universal moduli of continuity for solutions to fully nonlinear elliptic equations. Arch. Ration. Mech. Anal. **211**, 911–927 (2014)
35. E.V. Teixeira, Regularity for quasilinear equations on degenerate singular set. Math. Ann. **358**, 241–256 (2014)
36. E.V. Teixeira, Hessian continuity at degenerate points in nonvariational elliptic problems. Int. Math. Res. Not. IMRN **2015**(16), 6893–6906 (2015)
37. E.V. Teixeira, J.M. Urbano, A geometric tangential approach to sharp regularity for degenerate evolution equations. Anal. PDE **7**, 733–744 (2014)
38. J.M. Urbano, *The Method of Intrinsic Scaling*. Lecture Notes in Mathematics 1930 (Springer, Berlin, 2008)
39. J.L. Vázquez, *The Porous Medium Equation. Mathematical Theory*. Oxford Mathematical Monographs (Oxford University, Oxford, 2007)

Complete List of Mathematical Papers Authored by Emmanuele DiBenedetto

References

1. G. Alessandrini, E. DiBenedetto, Determining 2-dimensional cracks in 3–dimensional bodies: uniqueness and stability. Indiana Univ. Math. J. **46**(1), 1–82 (1997)
2. H.W. Alt, E. DiBenedetto, Flow of oil and water through porous media, in *Variational Methods for Equilibrium Problems of Fluids—Trento, 20–25 Juin 1983*. Astrisque, vol. 118 (1984), pp. 89–108
3. H.W. Alt, E. DiBenedetto, Non steady flow of water and oil through inhomogeneous porous media. Annali Scuola Norm. Sup. Pisa, Ser. IV **XII**(3), 335–392 (1985)
4. D. Andreucci, E. DiBenedetto, A new approach to initial traces in nonlinear filtration. Ann. Inst. H. Poincaré Anal. Non Linaire **7**(4), 305–334 (1990)
5. D. Andreucci, E. DiBenedetto, Behaviour at infinity of solutions of certain degenerate evolution equations with strongly nonlinear sources. Annali Scuola Norm. Sup. Pisa, Ser. IV **XV**(3), 363–441 (1991)
6. D. Andreucci, E. DiBenedetto, Weak solutions of equations of the type of nonlinear filtration. Non Linear Anal. TMA **19**(1), 29–41 (1992)
7. D. Andreucci, G. Caruso, E. DiBenedetto, Fingering in a Hele-Shaw cell as an obstacle-like problem. Indiana Univ. Math. J. **49**(4), 1287–1321 (2000)
8. D. Andreucci, P. Bisegna, E. DiBenedetto, Homogenization and concentrated capacity in reticular almost disconnected structures. C. R. Math. Acad. Sci. Paris **335**(4), 329–332 (2002)
9. D. Andreucci, P. Bisegna, E. DiBenedetto, Homogenization and concentrated capacity for the heat equation with non linear variational data in reticular almost disconnected structures, and applications to visual transduction. Ann. Mat. Pura Appl. **182**(4), 375–407 (2003)
10. D. Andreucci, P. Bisegna, G. Caruso, H. Hamm, E. DiBenedetto, Mathematical model of the spatio-temporal dynamics of second messengers in visual transduction. Biophysical J. **85**, 1358–1376 (2003)
11. D. Andreucci, G. Caruso, E. DiBenedetto, Ill-Posed Hele-Shaw flows, in *Free Boundary Problems*, vol. 147 (Birkhäuser, Boston, 2003), pp. 27–52
12. D. Andreucci, P. Bisegna, E. DiBenedetto, Mathematical problems in visual transduction, in *Trends in Partial Differential Equations in Mathematical Physics*, vol. 81 (Birkhäuser, Boston, 2005), pp. 65–80. Invited paper for the 70th birthday of V.A. Solonnikov
13. D. Andreucci, P. Bisegna, E. DiBenedetto, Homogenization and concentration of capacity in the rod outer segment with incisures. Appl. Anal. Special Issue Memory of C. Pucci **85**(1–3), 303–331 (2006)

V. Vespri et al. (eds.), *Harnack Inequalities and Nonlinear Operators*,
Springer INdAM Series 46, https://doi.org/10.1007/978-3-030-73778-8

14. T. Battacharya , E. DiBenedetto, J. Manfredi, Limits as $p \to \infty$ of $\Delta_p u_p = f$ and related extremal problems. Rend. Sem. Mat. Univ. Pol. Torino. (Fasc. Spec.) (1989), 15–68
15. G. Borgioli, E. DiBenedetto, M. Ughi, Stefan problems with nonlinear boundary conditions: the Polygonal method. ZAMM **58**, 530–546 (1978)
16. G. Borgioli, E. DiBenedetto, M. Ughi, Il Metodo di Hüber per problemi a contorno libero con dominio inizialmente degenere. Sem. Inst. Mat. U. Dini, Florence **10**, 1–23 (1978)
17. J.R. Cannon, E. DiBenedetto, On the existence of solutions of boundary value problems in fast chemical reactions. Boll. Un. Mat. Ital. **5**(15B), 835–843 (1978)
18. J.R. Cannon, E. DiBenedetto, The initial value problem for the Boussinesq equations with data in L^p, in *Approximation Methods for Navier Stokes Equations*. Springer Lecture Notes, vol. 771 (1979), pp. 129–144
19. J.R. Cannon, E. DiBenedetto, The steady state Stefan problem with convection with mixed temperature and nonlinear boundary conditions, in *Proceedings of Free Boundary Problems Symposium*, ed. by C. Baiocchi, E. Magenes. Pavia, (1979), pp. 232–265
20. J.R. Cannon, E. DiBenedetto, On the existence of weak solutions to an n-dimensional Stefan problem with nonlinear boundary conditions. SIAM J. Math. Anal. **11**(4), 632–645 (1980)
21. J.R. Cannon, E. DiBenedetto, G.K. Knightly, The steady state Stefan problem with convection. Arch. Rat. Mech. Anal. **73**, 79–97 (1980)
22. J.R. Cannon, E. DiBenedetto, G.K. Knightly, The bidimensional Stefan problem with convection: the time dependent case. Comm. Part. Diff. Equ. **8**(14), 1549–1604 (1983)
23. G. Caruso, H. Khanal, V. Alexiades, F. Rieke, H.E. Hamm, E. DiBenedetto, Mathematical and computational modeling of spatio-temporal signaling in rod phototransduction. IET Syst. Biol. **152**(3), 119–137 (2005)
24. G. Caruso, P. Bisegna, L. Shen, D. Andreucci, H.E. Hamm, E. DiBenedetto, Modeling the role of incisures in phototransduction in vertebrate photoreception. Biophys. J. **91**, 1192–1212 (2006)
25. G. Caruso, P. Bisegna, L. Shen, D. Andreucci, S. Gurevich, H.E. Hamm, E. DiBenedetto, Diffusion of the second messengers in the cytoplasm acts as a variability suppressor of the single photon response in vertebrate phototransduction. Biophys. J. **94**(9), 3363–3384 (2008)
26. G. Caruso, P. Bisegna, L. Lenoci, D. Andreucci, V. Gurevich, H.E. Hamm, E. DiBenedetto, Kinetics of Rhodopsin deactivation and its role in regulating recovery and reproducibility of rod photoresponse. PLoS Comput. Biol. **6**(12), 1–15 (2010)
27. G. Caruso, P. Bisegna, L. Lenoci, D. Andreucci, V. Gurevich, H.E. Hamm, E. DiBenedetto, Identification of key factors that reduce the variability of the single photon response. PNAS **108**(19), 7804–7807 (2011)
28. G.Q. Cheng, E. DiBenedetto, Stability of entropy solutions to the Cauchy problem for a class of hyperbolic-parabolic equations. SIAM J. Math. Anal. **33**(4), 751–762 (2001)
29. E. DiBenedetto, Regularity results for the Porous medium equation. Ann. Mat. Pura Appl. **CXXI**(IV), 249–262 (1980)
30. E. DiBenedetto, Regularity results for the two phase N-dimensional Stefan problem. Boll. Univ. Mat. Ital. **1**, 129–152 (1980)
31. E. DiBenedetto, Interior and boundary regularity for a class of free boundary problems. Pitman **78**. Edited by A. Fasano, M. Primicerio (1981), 383–395
32. E. DiBenedetto, Continuity of weak solutions to certain singular Parabolic equations. Ann. Mat. Pura Appl. **CXXX**(IV), 131–177 (1982)
33. E. DiBenedetto, Continuity for singular Parabolic equations, in *Nonlinear Phenomena in Mathematical Sciences*, ed. by Laksmikantam (Acadamic Press, New York, 1982), pp. 307–312
34. E. DiBenedetto, Continuity of weak solutions to a general porous medium equation. Indiana Univ. Math. J. **32**(1), 83–118 (1983)
35. E. DiBenedetto, Local regularity of weak solutions of degenerate elliptic equations. Non Linear Anal. TMA **7**(8), 827–850 (1983)

36. E. DiBenedetto, Change of phase with convection, in *Applied Nonlinear Functional Analysis, Peter Language*, ed. by R. Gorenflo, K.H. Hoffmann, vol. 25 (1983), pp. 1–18
37. E. DiBenedetto, Conduction-convection problems with change of phase, in *Free Boundary Problems and Applications, Pitman*, ed. by Fremond, Damlamian, Bossavit (1984), pp. 1–14
38. E. DiBenedetto, The flow of two immiscible fluids through a porous medium; regularity of the saturation, in *Theory and Applications of Liquid Crystals, Springer Verlag Lecture Notes*, vol. 5, ed. by D. Kinderlehrer, N. Ericksen (1985), pp. 123–143
39. E. DiBenedetto, Periodic solutions of the Dam problem, nonlinear evolution equations: qualitative properties of solutions. Pitman **149**, 77–84 (1985). Edited by A. Tesei, L. Boccardo
40. E. DiBenedetto, *Remarks on Some Ill Posed Problems*. Nonlinear PDE's, ed. by N. Trudinger, L. Simon, (Australian National University, Camberra, 1985), pp. 32–56
41. E. DiBenedetto, On the local behavior of solutions of degenerate parabolic equations with measurable coefficients. Ann. Scuola Norm. Sup. Pisa, Ser. IV **X**(3), 487–535 (1986)
42. E. DiBenedetto, Harnack type inequalities for some degenerate Parabolic equations, in *Nonlinear Diffusion Equations and their Equilibrium States Mathematical Sciences Research Institute*, vol. 12, ed. by W.M. Ni, L.A. Peletier, J. Serrin (Springer, New York, 1987), pp. 267–272
43. E. DiBenedetto, Intrinsic Harnack–type inequalities, for solutions of degenerate Parabolic equations. Arch. Rat. Mech. Anal. **100**(2), 129–147 (1987)
44. E. DiBenedetto, Local regularity and extinction free boundary for a singular Parabolic equation, in *Free Boundary Problems and Applications, Pitman*, ed. by J. Sprekels, K.H. Hoffman (1988), pp. 1–7
45. E. DiBenedetto, Initial traces for degenerate and singular evolution equations, problems involving change of type, in *Springer Lecture notes in Physics*, vol. 359. Conference for the 60th birthday of J. Hale, ed. by K. Kirchgässner (1988), pp. 175–190
46. E. DiBenedetto, Harnack estimates in certain function classes. Atti Sem. Mat. Fis. Univ. Modena **37**, 173–182 (1989)
47. E. DiBenedetto, Degenerate and singular parabolic equations, in *Recent Advances in Partial Differential Equation, John Wiley, Research in Applied Mathematics*, ed. by P.G. Ciarlet (1994), pp. 55–84
48. E. DiBenedetto, Parabolic equations with multiple singularities or degeneracies. Equadiff **9**(Brno), 25–48 (1997)
49. E. DiBenedetto, D. Diller, About a singular parabolic equation arising in thin film dynamics and in the ricci flow for complete R^2, in *Lecture Notes in Pure and Applied Mathematics*, vol. **177**. Special Volume for the 70th birthday of C. Pucci. Edited by G. Talenti, E. Vesentini, P. Marcellini (Marcel Dekker, New York, 1995), pp. 103–120
50. E. DiBenedetto, D. Diller, On the rate of drying of gelatin in a photographic film. Adv. Differ. Equ. **1**(6), 989–1003 (1996)
51. E. DiBenedetto, D. Diller, A new form of the Sobolev multiplicative inequality and applications to the asymptotic decay of solutions to the Neuman problem for quasilinear parabolic equations with measurable coefficients. Nat. Acad. Sci. Ukraine Nonlinear Boundary Value Prob. Kiev Ukraine **8**, 81–88 (1997)
52. E. DiBenedetto, D. Diller, Singular semilinear elliptic equations in $L^1(R^N)$, in *Non Linear Differential Equations and their Applications*, vol. 35. Invited paper for the H. Amann Anniversary Volume (Birkhäuser, Boston, 1998), pp. 143–182
53. E. DiBenedetto, C. Elliott, Existence for a problem in ground freezing. Non Linear Anal. TMA **9**(9), 953–968 (1985)
54. E. DiBenedetto, A. Friedman, The Ill-Posed Hele-Shaw model and the Stefan problem for supercooled water. Trans. Am. Math. Soc. **282**(1), 183–204 (1984)
55. E. DiBenedetto, A. Friedman, Regularity of solutions of nonlinear degenerate parabolic systems. J. Reine Angew. Math. **349**, 83–128 (1984)
56. E. DiBenedetto, A. Friedman, Hölder estimates for nonlinear degenerate parabolic systems. J. Reine Angew. Math. **357**, 1–22 (1985)

57. E. DiBenedetto, A. Friedman, Bubble growth in porous media. Indiana Univ. Math. J. **35**(3), 573–606 (1986)
58. E. DiBenedetto, A. Friedman, A boundary modulus of continuity for a class of singular Parabolic equations. J. Diff. Equ. **63**(3), 418–447 (1986)
59. E. DiBenedetto, A. Friedman, Periodic behavior of solutions for the evolutionary Dam problem and related free boundary problems. Comm. Part. Differ. Equ. **11**(12), 1297–1377 (1986)
60. E. DiBenedetto, A. Friedman, Conduction-convection with a change of phase. J. Differ. Equ. **62**(2), 129–185 (1986)
61. E. DiBenedetto, R. Gariepy, On the local behavior of solutions for an Elliptic-Parabolic equation. Arch. Rat. Mech. Anal. **97**(1), 1–17 (1987)
62. E. DiBenedetto, U. Gianazza, A wiener-type condition for boundary continuity of Quasi-Minima of variational integrals. Manuscripta Math. **149**(3–4), 339–346 (2016)
63. E. DiBenedetto, U. Gianazza, Some properties of the DeGiorgi classes. Rend. Istit. Mat. Univ. Trieste **48**, 245–260 (2016)
64. E. DiBenedetto, M.A. Herrero, On the Cauchy problem and initial traces for a degenerate parabolic equation. Trans. Am. Math. Soc. **314**(1), 187–224 (1989)
65. E. DiBenedetto, M.A. Herrero, Nonnegative solutions of the evolution p-laplacian equation; initial traces and Cauchy problem when $1 < p < 2$. Arch. Rat. Mech. Anal. **1**(3), 225–290 (1990)
66. E. DiBenedetto, D. Hoff, An interface tracking algorithm for the porous medium equation. Trans. Am. Math. Soc. **284**(2), 463–500 (1984)
67. E. DiBenedetto, Y.C. Kwong, Harnack estimates and extinction profile for weak solutions of certain singular parabolic equations. Trans. Am. Math. Soc. **330**(2), 783–811 (1992)
68. E. DiBenedetto, J. Manfredi, On the higher integrability of the gradient of weak solutions of certain degenerate elliptic systems. Am. J. Math. **115**(5), 1107–1134 (1993)
69. E. DiBenedetto, M. O'Leary, Three-dimensional conduction-convection with change of phase. Arch. Rat. Mech. Anal. **123**, 99–116 (1993)
70. E. DiBenedetto, M. O'Leary, Local C^{α}Convergence of elliptic regularizations of parabolic equations. Meccanica **28**, 97–102 (1993)
71. E. DiBenedetto, M. O'Leary, Conduction-convection problems with change of phase, in *Nonlinear Problems in Applied Mathematics, SIAM Special Volume, for the 70th Birthday of I. Stackgold*, ed. by T.S. Angell, L.P. Cook, R.E. Kleinman, W.E. Olmstead (1995), pp. 104–115
72. E. DiBenedetto, M. Pierre, On the maximum principle for Pseudo-Parabolic equations. Indiana Univ. Math. J. **30**, 821–854 (1981)
73. E. DiBenedetto, R.E. Showalter, Implicit degenerate evolution equations and applications. SIAM J. Math. Anal. **12**(5), 731–751 (1981)
74. E. DiBenedetto, R.E. Showalter, A Pseudo-Parabolic variational inequality and Stefan problem. Non Linear Anal. TMA **6**, 279–299 (1982)
75. E. DiBenedetto, R.E. Showalter, Free boundary problems for a degenerate parabolic system. J. Diff. Equ. **50**(1), 1–19 (1983)
76. E. DiBenedetto, R. Spigler, An algorithm for the one phase Stefan problem. Rend. Sem. Mat. Univ. Padova **69**, 109–134 (1983)
77. E. DiBenedetto, N.S. Trudinger, Harnack inequalities for Quasi-Minima of variational integrals. Ann. Inst. H. Poincaré Anal. Non Linaire **1**(4), 295–308 (1984)
78. E. DiBenedetto, V. Vespri, Continuity for bounded solutions of multiphase Stefan problem. Atti Accad. Naz. Lincei Cl. Sci. Fis. Mat. Natur. Rend. Lincei (9) Mat. Appl. **5**(4), 297–302 (1994)
79. E. DiBenedetto, V. Vespri, On the singular equation $b(u)_t = \Delta u$. Arch. Rat. Mech. Anal. **132**, 247–309 (1995)
80. E. DiBenedetto, A. Fasano, M. Primicerio, On a free boundary problem related to an irreversible process. Control Cybern. **14**(1–3), (1985), 195–219

81. E. DiBenedetto, C. Elliot, A. Friedman, The free boundary of a flow in a porous body heated from its boundary. Non Linear Anal. TMA **10**, 879–900 (1986)
82. E. DiBenedetto, Y.C. Kwong, V. Vespri Local analiticity and asymptotic behaviour of solutions of certain singular parabolic equations. Indiana Univ. Math. J. **40**(2), 741–765 (1991)
83. E. DiBenedetto, J. Manfredi, V. Vespri, A note on boundary regularity for certain degenerate parabolic equations, in *Nonlinear Diffusion Equations and Their Equilibrium States*, ed. by N.G. Lloyd, W.M. Ni, L.A. Peletier, J. Serrin (Birkhäuser, Boston, 1992), pp. 177–182
84. E. DiBenedetto, D. Diller, S. Davis, Some a priori estimates for a singular evolution equation arising in thin film dynamics. SIAM J. Math. Anal. **27**(3), 638–660 (1996)
85. E. DiBenedetto, J.M. Urbano, V. Vespri, Current issues on singular and degenerate evolution equations, in *Invited Chapter for the Handbook of Differential Equations Elsevier*, ed. by C. Dafermos, E. Feireisl, vol. 1 (2004), pp. 169–286
86. E. DiBenedetto, U. Gianazza, V. Vespri, Local clustering of the non-zero set of functions in $W^{1,1}$. Atti Accad. Naz. Lincei Rend. Lincei Mat. Appl. **17**(3), 223–225 (2006)
87. E. DiBenedetto, U. Gianazza, V. Vespri, Intrinsic Harnack estimates for nonnegative local solutions of degenerate parabolic equations. ERA-AMS **12**, 95–99 (2006)
88. E. DiBenedetto, U. Gianazza, V. Vespri, Intrinsic harnack estimates for non-negative solutions of quasilinear singular parabolic partial differential equations. Rend. Accad. Naz. Lincei **18**, 359–364 (2007)
89. E. DiBenedetto, U. Gianazza, V. Vespri, Intrinsic harnack estimates for non-negative solutions of quasilinear degenerate parabolic equations. Acta Math. **200**(2), 181–209 (2008)
90. E. DiBenedetto, U. Gianazza, V. Vespri, Sub-Potential lower bounds for non-negative solutions to certain quasi-linear degenerate parabolic equations. Duke Math. J. **143**, 1–15 (2008)
91. E. DiBenedetto, U. Gianazza, V. Vespri, Alternative forms of the harnack inequality for non-negative solutions to certain degenerate and singular parabolic equations. Atti Accad. Naz. Lincei Rend. Lincei Mat. Appl. **20**, 369–377 (2009)
92. E. DiBenedetto, U. Gianazza, V. Vespri, Forward, backward and elliptic harnack inequalities for non-negative solutions to certain singular parabolic differential equations. Ann. Scuola Norm. Sup. Pisa Cl. Sci. (5) **9**(2), 385–422 (2010)
93. E. DiBenedetto, U. Gianazza, V. Vespri, Continuity of the saturation in the flow of two immiscible fluids in a Porous medium. Indiana Univ. Math. J. **59**, 2041–2076 (2010)
94. E. DiBenedetto, U. Gianazza, V. Vespri, Harnack type estimates and holder continuity for non-negative solutions to certain sub-critically singular parabolic partial differential equations. Manuscripta Math. **131**, 231–245 (2010)
95. E. DiBenedetto, U. Gianazza, V. Vespri, A new approach to the expansion of the positivity set of non-negative solutions to certain singular parabolic partial differential equations. Proc. Am. Math. Soc. **138**(10), 3521–3529 (2010)
96. E. DiBenedetto, U. Gianazza, V. Vespri, Liouville-type theorems for certain degenerate and singular parabolic equations. C. R. Math. Acad. Sci. Paris **348**, 873–877 (2010)
97. E. DiBenedetto, U. Gianazza, N. Liao, On the local behavior of a logarithmically singular equation. Discrete Contin. Dyn. Syst. Ser. B **17**(6), 1841–1858 (2012)
98. E. DiBenedetto, U. Gianazza, N. Liao, Logarithmically singular parabolic equations as limits of the Porous medium equation. Non Linear Anal. TMA **85**(12), 4513–4533 (2012)
99. E. DiBenedetto, U. Gianazza, N. Liao, *Two remarks on the local behavior of solutions to logarithmically singular diffusion equations and its porous-medium type approximations*. Riv. Math. Univ. Parma (N.S.) **5**(1), 139–182 (2014)
100. E. DiBenedetto, U. Gianazza, C. Klaus, A necessary and sufficient condition for the continuity of local minima of parabolic variational integrals with linear growth. Adv. Calc. Var. **10**(3), 209–221 (2017)
101. E. DiBenedetto, U. Gianazza, V. Vespri, Remarks on local boundedness and local Hölder continuity of local weak solutions to anisotropic p-Laplacian equations. J. Elliptic Parabol. Equ. **2**(1), 157–169 (2017)

102. P. Hinow, C.E. Rogers, C.E. Barbieri, J.A. Pietenpol, A. Kenworthy, E. DiBenedetto, The DNA binding activity of p53 displays reaction-diffusion kinetics. Biophys. J. **91**, 330–342 (2006)
103. M. Kang, C.A. Day, K.R. Drake, A. Kenworthy, E. DiBenedetto, A generalization of theory for 2D fluorescence recovery after photobleaching, applicable to confocal laser scanning microscopes. Biophys. J. **97**, 1501–1511 (2009)
104. M. Kang, C.A. Day, E. DiBenedetto, A.K. Kenworthy, A quantitative approach to analyze binding diffusion kinetics by confocal FRAP. Biophys. J. **99**(5), 2737–2747 (2010)
105. M. Kang, E. DiBenedetto, A.K. Kenworthy, Proposed correction to the Feder's anomalous diffusion FRAP equations. Biophys. J. **100**(3), 791–792 (2011)
106. M. Kang, C.A. Day, A.K. Kenworthy, E. DiBenedetto, Simplified equation to extract diffusion coefficients from confocal FRAP data. Traffic **13**, 1589–1600 (2012)
107. H. Khanal, V. Alexiades, E. DiBenedetto, H. Hamm, Numerical simulations of diffusion of second messengers cGMP and Ca2+ in rod photoreceptors, in *American Institute of Physics*, vol. 665. Unsolved Problems of Noise and Fluctuations, Washington DC (2003), pp. 165–172
108. H. Khanal, V. Alexiades, E. DiBenedetto, Response of dark-adapted retinal rod photoreceptors, in *Proceedings of Dynamical Systems and Applications, Morehouse*, vol. 4 (2004), pp. 138–145
109. C. Klaus, K. Raghunathan, E. DiBenedetto, A.K. Kenworty, *Analysis of diffusion in curved surfaces and its application to tubular membranes*. Mol. Biol. Cell **27**, 3937–3946 (2016)
110. C. Klaus, G. Caruso, V.V. Gurevich, E. Di Benedetto, Multi-scale, numerical modeling of spatio-temporal signaling in cone phototransduction. PLoS One **14**(7), e0219848 (2019)
111. L. Lenoci, M. Duvernay, S. Satchell, E. DiBenedetto, H.E. Hamm, Mathematical model of PAR1-mediated activation of platelets. Mol. Biosyst. **7**(4), 1129–1137 (2011)
112. L. Lenoci, H.E. Hamm, E. DiBenedetto, Identification of the key parameters in a mathematical model of PAR1-mediated signaling in endothelial cells. arXiv:1101.3215v1 [q-bio.MN] (2011)
113. C.L. Makino, X.-H. Wen, N.A. Michaud, H.I. Covington, E. DiBenedetto, H.E. Hamm, J. Lem, G. Caruso, Rhodopsin expression level affects Rod outer segment morphology and photoresponse kinetics. PLoS One **7**(7), 1–7 (2012)
114. J.N. McLaughlin, L. Shen, M. Holinstat, J.D. Brooks, E. DiBenedetto, H. Hamm, Functional selectivity of G protein signaling by agonist peptides and Thrombin for the protease-activated receptor-1. J. Biol. Chem. Mol. Biol. **280**(6), 25048–25059 (2005)
115. L. Shen, D. Andreucci, H.E. Hamm, E. DiBenedetto, Fluctuations of the single photon response in visual transduction. Am. Inst. Phys. Noise Fluctuations **780**, 553–556 (2005)
116. L. Shen, P. Bisegna, G. Caruso, D. Andreucci, S. Gurevich, H.E. Hamm, E. DiBenedetto, Dynamics of mouse rod phototransduction and its sensitivity to variation of key parameters. IET Syst. Biol. **4**(1), 12–32 (2010)
117. X.-H. Wen, L. Shen, R.S. Brush, N. Michaud, M.R. Al-Ubaidi, V. Gurevich, H.E. Hamm, J. Lem, E. DiBenedetto, R.E. Anderson, C.L. Makino, Over-expression of rhodopsin alters the structure and photoresponse of rod photoreceptors. Biophys. J. **96**(3), 929–950 (2009)
118. C. Ya-zhe, E. DiBenedetto, On the local behaviour of solutions of singular Parabolic equations. Arch. Rat. Mech. Anal. **103**(4), 319–346 (1988)
119. C. Ya-zhe, E. DiBenedetto, Boundary estimates for solutions of nonlinear degenerate parabolic systems. J. Reine Angew. Math. **395**, 102–131 (1989)
120. C. Ya-zhe, E. DiBenedetto Hölder estimates of solutions of singular parabolic equations with measurable coefficients. Arch. Rat. Mech. Anal. **118**, 257–271 (1992)
121. C. Ya-zhe, E. DiBenedetto, On the harnack inequality for nonnegative solutions of singular parabolic equations, in *Degenerate Diffusion; The IMA Volumes in Math. and its Application*, vol. 47, ed. by W.M. Ni, L.A. Peletier, J.L. Vazquez (Springer, Berlin, 1993), pp. 61–69

Complete List of Monographs and Textbooks Authored or Edited by Emmanuele DiBenedetto

122. G.Q. Cheng, E. DiBenedetto, Nonlinear partial differential equations, in *AMS, Contemporary Mathematics*, vol. 238 (AMS, Providence, 1999)
123. E. DiBenedetto, *Fully Non Linear Elliptic and Parabolic Equations*. Lecture Notes Northwestern University (1986)
124. E. DiBenedetto, *Quasilinear Hyperbolic Systems in $\mathbb{R}^N$ With Smooth Data*. Lecture Notes Northwestern University (1990)
125. E. DiBenedetto, Degenerate and singular parabolic equations in divergence form and with measurable coefficients, in *Lipschitz Lectures, Institute für Angewandte Mathematical* (Bonn, Germany, 1992)
126. E. DiBenedetto, *Degenerate Parabolic Equations* (Springer, New York, 1993)
127. E. DiBenedetto, *Partial Differential Equations* (Birkhäuser, Boston, 1995)
128. E. DiBenedetto, *Real Analysis*. Advanced Text Series (Birkhäuser, Boston, 2002)
129. E. DiBenedetto, *Real Analysis, Chinese Ed.* (Chinese Higher Education Press, China, 2007)
130. E. DiBenedetto, *Partial Differential Equations*, 2nd edn. (Birkhäuser, Boston, 2009)
131. E. DiBenedetto, *Classical Mechanics* (Birkhäuser, Boston, 2010)
132. E. DiBenedetto, *Real Analysis*, 2nd edn. Advanced Text Series (Birkhäuser, Boston, 2016)
133. E. DiBenedetto, U. Gianazza, V. Vespri, Harnack inequality for degenerate and singular parabolic equations, in *Springer Monographs in Mathematics* (2011)

V. Vespri et al. (eds.), *Harnack Inequalities and Nonlinear Operators*,
Springer INdAM Series 46, https://doi.org/10.1007/978-3-030-73778-8